10/92

BLACK HOLES IN SPACETIME

BLACK HOLES IN SPACETIME

BY KITTY FERGUSON

A VENTURE BOOK
FRANKLIN WATTS   1991
NEW YORK/LONDON/TORONTO/SYDNEY

On the cover: Computer simulation by Emilio Falco of the Harvard Center for Astrophysics showing a galaxy as it might appear if a black hole were to come between it and us. The paths of light coming from some of the stars in the galaxy are bent around all sides of the black hole at once. They reach our eyes as a bright ring, known as an Einstein ring.

Photographs courtesy of David F. Malin: p. 14; Lick Observatory: p. 19; University of Chicago Press/S. Chandrasekhar: p. 22; Smithsonian Astrophysical Laboratory/Emilio E. Falco: pp. 72, 73; Princeton University/ Robert P. Matthews: p. 75; Miriam Berkley: p. 79; Harvard-Smithsonian Center for Astrophysics: pp. 95, 98; Lola Chaisson: pp. 97, 107; Harvard College Observatory: p. 105; National Radio Astronomy Observatory/AUI: p. 113.

Library of Congress Cataloging-in-Publication Data

Ferguson, Kitty.
Black holes in spacetime / by Kitty Ferguson.
p.   cm. — (A Venture book)
Includes bibliographical references and index.
Summary: Discusses the formation and possible behavior of black holes and why we believe that they exist.
ISBN 0-531-12524-6
1. Black holes (Astronomy)—Juvenile literature.   [1. Black holes (Astronomy)]   I. Title.
QB843.B55F47   1991
523.8′875—dc20
91-2111 CIP AC

*To my daughter, Caitlin,
whose third-grade science fair project
on black holes inspired this book*

The author would like to express her thanks to:

John A. Wheeler for his kind help and encouragement.

*Yale Ferguson, Matthew Fremont, Stephen Hawking,
Howard Helms,
William Unruh, David and Judy Vetter,
Herman and Tina Vetter, and Patrick Yates,
for reading and checking over portions of this book.*

*David Arnett, Eric Chaisson, Nicholas Phillips, Fay
Dowker, Andrew Dunn, Sir Brian Pippard,
Stuart Shapiro, and Saul Teukolsky,
for answering many detailed questions.*

*None of the above should be held accountable for any
mistakes in this book.
The author assumes full responsibility.*

CONTENTS

Prologue
11

1
Burnout on the Cosmic Level:
The Life Cycle of a Star
13

2
The Awesome Power of Gravity
28

3
Treacherous Voyage
43

4
Contemplating an Enormous Nothing
69

5
Evidence in the Case for Black Holes
90

6
Wild Ideas
116

7
Gravity Is Patient
124

Glossary   127

Sources Used   135

Index   139

BLACK HOLES IN SPACETIME

PROLOGUE

For hundreds of thousands of years, people living on the earth have watched the night sky, awed by its mysteries. We've yearned to understand what we see there . . . and what we can't see.

Until 1609 we had only our eyes, our mathematics, and our imaginations to help us. But in that year the first telescopes appeared and opened up vast new possibilities. We've steadily improved our telescopes and our understanding. By the 1930s we were using radio receivers to discover things in space that aren't visible through any optical telescope.

Even then, there was a lot we were missing. Our earth's atmosphere is a barrier. We were like crabs scuttling on the ocean floor, wondering what we could see if we stuck our heads above water. We put telescopes in orbit beyond our atmosphere, and we found much more in the heavens than we could see on a clear night with even the most powerful earthbound telescopes.

Much more . . . but not everything we know is there.

No one has seen a **black hole.**

A few years ago the question of whether there really are any black holes was still unanswered. Today, not so. We have good evidence of the presence of several black holes and reasons to think there are many more. The evidence is indirect—circumstantial, it would be called in a court of law—but convincing.

Have black holes become a ho-hum commonplace thing—"Okay, we've found that; let's get on to something else"? Far from it.

Within black holes lie clues to what physicist John Wheeler calls the "deep, happy mysteries" of the universe. The attempt to unlock these mysteries is the greatest adventure in modern science. But black holes don't give up their secrets easily. Finding black holes hasn't allowed us to probe inside them. We may never be able to do that.

What are these secretive invisible objects? Why are they there? What makes us so certain they *are* there, if we can't see them?

1
BURNOUT ON THE COSMIC LEVEL: THE LIFE CYCLE OF A STAR

It is therefore possible that the greatest luminous bodies in the universe are on this very account invisible.
—*Pierre-Simon Laplace, 1795*

Before you can begin to think intelligently about black holes and understand why we're certain they exist, you need to know something about stars.

Stars don't last forever. They're born, and shine for millions or billions of years, and die. A black hole is one of the ways a star can spend its old age. It is a "final state" of a star.

THE BIRTH OF A STAR

Imagine an enormous cloud of gas out in space, tons and tons of it. There are theories that try to explain why more gas (or anything else) has accumulated in one part of the universe than another. Whatever the cause, it's obvious such "clumping" has happened.

13

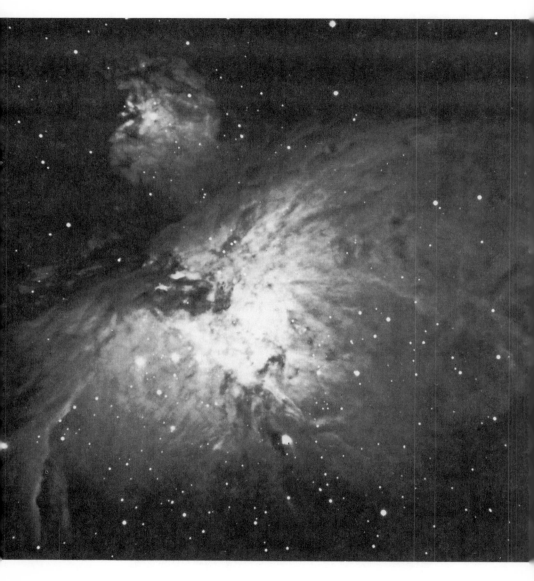

The gaseous Great Nebula in Orion is a region of star formation.

The gas is mostly **hydrogen**. This hydrogen gas, like all other ordinary matter in the universe, is made up of **atoms.** We've all seen textbook pictures of atoms: a nucleus, made up of protons and neutrons with electrons in orbit around it. These pictures don't even begin to tell you how much empty space is in an atom. You'll be near the right idea if you imagine that the nucleus is a baseball and the electrons are Ping-Pong balls about three miles away!

The atoms and particles in the hydrogen gas we're imagining have come close enough together to begin pulling on one another and drawing closer together still. This pulling, of course, is what we call **gravity,** or gravitational attraction. We experience it all the time on the earth. Sometimes we forget that it's also at work among the smallest particles and among all the enormous objects out in space.

As the gravity among the atoms in the gas draws these atoms closer together, they bump into each other. Whenever objects come nearer to one another, the gravitational attraction between them gets stronger. This causes the atoms to draw closer yet, moving faster because of the stronger gravitational pull. Collisions happen more and more often and at higher speeds. Everything heats up.

When things get hot enough, something new begins to occur. The **hydrogen atoms** in the gas don't bounce off one another anymore. Their **nuclei** start to merge into one another and change into **helium**.

This transformation from hydrogen to helium is what we call **nuclear fusion.** Because of the overcrowded conditions among the atoms in the gas, there will be a lot of it going on. In fact, it's like a controlled hydrogen bomb explosion, which causes an enormous amount of heat. The gas has pulled itself together to form a star. The release of energy makes the star shine.

THE ADULT YEARS
OF A STAR

Stars don't last forever, but they do last a long time, not getting any smaller for a while. Why not? Won't the star keep shrinking just the way the cloud of gas did to form it? Aren't the atoms going to keep on pulling at one another and getting closer together? Of course they are, with more and more nuclear fusion, more and more hydrogen changing to helium and from helium to heavier elements, more and more heat.

In fact, the heat from nuclear reactions increases the pressure of the gas. This pressure fights back against gravity (Figure 1). It keeps the star from collapsing. Think of the pressure in a balloon that holds the sides of the balloon apart. The rubber sides of the balloon are pulling toward each other, trying to come together. You find that out quickly enough if you let the gas out. But the pressure of the gas inside the balloon won't let the sides come together. The pressure in the star won't let it collapse.

For a while, we have a nice balance—a tug-of-war between two teams. Pressure from the heat of nuclear reactions on one team, gravity on the other team (Figure 1).

Occasionally gravity seems to be winning. The accumulation of helium ash may snuff out the central furnace of the star. The core starts to collapse. But soon the collapse again produces high temperatures and the star reignites (now fusing helium into carbon atoms). In stars much more massive than our sun, the snuffing out and reignition might reoccur several times as the star fuses its atoms into heavier and heavier elements. As we've said, the teams (gravity, and pressure from nuclear reactions) are closely matched, and the competition won't end for a long

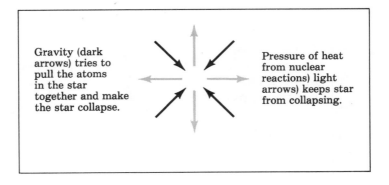

Figure 1. Gravity vs. nuclear reactions

time. The lifetime of a star like the sun is 10 or 11 billion years.

The game eventually ends, and when it does, gravity wins. One of the things we notice in the universe is that gravity is patient. Patience pays off. At last most of the **hydrogen atoms** have changed into **helium atoms** and some of those into carbon atoms and, in the largest stars, into heavier elements. The star finally runs out of usable fuel. There are fewer and fewer nuclear reactions, and less and less heat and pressure to counteract the gravitational attraction. After hanging in there for millions or even billions of years, gravity wins the tug-of-war. The cooling star begins to shrink and collapse.

THE OLD AGE
OF A STAR

Will the star just fizzle out? What can stop it? There are now three things that may happen to the star.

1. It may stop shrinking and settle down to spend its old age as a **white dwarf,** a few thousand miles across, not too much smaller than the earth, with a density of hundreds of tons per cubic inch. Sirius, the brightest star in our night sky, has a white dwarf orbiting it.

2. It may shrink smaller than that until it's only about 20 miles (30 kilometers) across, then stop shrinking and become a **neutron star,** with an unbelievable density of millions of tons per cubic inch. **Pulsars,** for instance, are neutron stars that rotate many times a second and send out regular pulses of radio waves and other types of radiation.

3. It may keep on shrinking until it's nothing but a point in space where the density is infinite; in other words, it has formed a black hole.

First, let's find out why some stars *don't* shrink all the way down to form black holes. What stops them? It can't be the pressure of heat from nuclear reactions anymore. All the fuel for that is used up. These stars are cold stars.

Gravity, trying to collapse the star, has a new opponent now, after nuclear reactions have left the playing field; that opponent is something we call the **exclusion principle.**

Remember that atoms are not the smallest things in the universe. They are made up of much tinier building blocks called particles, and some of these particles are separated by a great amount of empty space. (If you think your desk, for instance, is "solid," you're mistaken.) All the particles of matter—of atoms, of you, of this book, of a star—obey the exclusion principle.

18

*Sirius B, a white dwarf, seen here
as a small dot of light. It orbits
Sirius A (the Dog Star), the brightest
star in our night sky. The "spikes" in
the picture aren't real—they are caused
by the support struts of the telescope.*

Here's the rule: Two particles of matter can't occupy precisely the same **quantum state.** What does that mean? It means that two particles that are identical—let's say, for instance, two **electrons,** with the same mass, spin, and electric charge—can't be in the same position *and* have the same velocity. To put that more simply and in a way most useful to us: *If two of these similar particles are very nearly in the same position, they must have different velocities.* And they won't stay in the same position long.

The exclusion principle sounds unlikely, but it's only one of many peculiar rules we discover in nature, rules that cause our universe to exist in the way we take for granted. The exclusion principle makes certain that particles keep their distance. It makes certain there will be empty space within atoms. Without it we wouldn't have chairs and people and stars. We'd have soup.

It's obvious that as the star shrinks, the particles become more tightly packed together. The more tightly they're packed together, the stronger the pull of gravity becomes among them. It gets more and more difficult for the particles to obey the exclusion principle. Remember, if they are very nearly in the same positions, they *must* have different velocities. This makes them move away from one another, opposing gravity.

Will gravity be able to pull the particles so tightly that the exclusion principle will be overpowered? In some stars, yes. In others, no.

In less-massive stars we have the new tug-of-war between closely matched teams: gravity versus the exclusion principle (Figure 2). The exclusion principle stops the collapse of the star. It won't let the particles come any closer together. We get a white dwarf

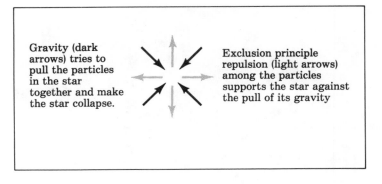

Gravity (dark arrows) tries to pull the particles in the star together and make the star collapse.

Exclusion principle repulsion (light arrows) among the particles supports the star against the pull of its gravity

Figure 2. Gravity vs. the exclusion principle

or a neutron star—small, but not so small as a black hole.

In stars with more mass, gravity will overpower the exclusion principle, and the star will continue to shrink.

Keep in mind as we go on that mass is not the same thing as size. A small object can be massive. Within the earth's gravity, this is the same as saying that a small object can be heavy, and we all know that's true. We can say that **mass** is a measure of how much matter is in the object (regardless of how closely the matter is packed) and how much the object resists any attempt to change its speed or direction. The important thing, for this book, is to understand that a shrunken star—a white dwarf or a neutron star or a black hole—can still have just as much mass as it did before it shrank, even though it has become much, much smaller.

A very massive star will shrink to form a black hole, but just how massive is "very" massive?

21

Subrahmanyan Chandrasekhar, the Indian physicist who calculated that a cold star with more than one and a half times the mass of our sun wouldn't be able to support itself against the pull of gravity. This is known as the "Chandrasekhar limit."

Subrahmanyan Chandrasekhar calculated in the late 1920s that if a star's mass is more than about one and a half times the mass of our sun, the exclusion principle will not be able to hold its own in the tug-of-war with gravity. We call this mass the **Chandrasekhar limit.** The force of gravity in a star more massive than the Chandrasekhar limit will overpower the exclusion principle repulsion among the electrons. The star won't end up as a white dwarf. It will continue to collapse (Figure 3).

We now figure that if the mass of a star (or the mass of what remains after a star explodes) is less than two or three solar masses (two or three times the mass of our sun), even though the exclusion principle repulsion among the electrons is overpowered, there may not be enough gravity to overpower the exclusion principle repulsion among the **neutrons.** A star like that will shrink to a neutron star. Stars more massive than two or three solar masses will probably collapse to black holes.

Some stars that are above the limit explode and throw off some of their matter. We call this explosion a **supernova.** After a supernova, a star may still have too much mass to become anything except a black hole. But a supernova can save a star from becoming a black hole and cause it to become a neutron star instead, if the star loses enough mass. You will read in some books that a star must supernova on the way to becoming a black hole. Most experts, however, believe that a star can become a black hole without ever exploding, particularly if the star is extremely massive.

It may seem odd to you that it's the *more* massive stars which end up as black holes, while *less* massive stars end up as something larger, white dwarfs and neutron stars. Common sense says it should be the other way around.

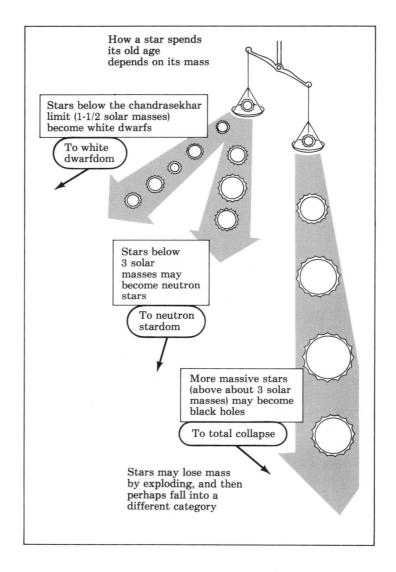

Figure 3. How a star spends its old age depends on its mass.

Remember that it's the gravitational attraction among the matter particles that's making the star collapse. The more massive the star, the more particles of matter there are in it. The more particles of matter, the more gravitational attraction. The more massive the star, the more powerful the gravity team—and the result is a big problem for the exclusion principle team.

The problem is an awful lot of particles are being crowded close together. They have to find more and more different velocities, so that no two which are close together will have the same velocity. The average velocity becomes higher and higher.

There's a limit to this—an unbreakable speed limit in the universe. Albert Einstein's equations show us that nothing in the universe can move faster than the **speed of light.** That speed is now figured at about 186,000 miles (300,000 kilometers) per second. When gravity in a star becomes so enormously strong that **matter** particles can only obey the exclusion principle by exceeding the speed of light (which they can't do), the exclusion principle is overpowered. Gravity wins again (Figure 4).

After that the star's fate is sealed. Gravity, the same force that gave birth to the star millions or billions of years before, finally claims the star as its victim. It crushes all the enormous mass of the star to the size of our sun . . . then to the size of the earth . . . to the size of a city . . . of your house . . . of you . . . a ball . . . a marble . . . a pinhead . . . a microbe. . . . Nothing can stop the collapse now, not even after the star becomes a black hole.

In one way of thinking about it, a black hole isn't a star, it's what happens in **spacetime** *around* a star that goes on shrinking to infinite density.

Electrons and other particles of matter avoid close company. They have a rule, called the EXCLUSION PRINCIPLE, which demands that *when particles are alike* in spin, mass, and charge, *and are also in the same position, they can't have the same velocity*. This ensures that they won't stay in the same position long.

GRAVITY on the other hand opposes this snobbishness among particles, and it enforces some rules of its own. For instance, the more matter particles there are in an object, each with its own gravitational attraction, the more gravity there is, and the harder it is for the particles to keep their distances. Also, as the particles get closer, they feel one another's gravity more strongly.

The only solution for the partcles is to move at more and more different velocities. For some of them this will mean becoming increasingly energetic. It's like a skating rink where the skaters all detest one another and are trying to avoid any contact, but at the same time have a force among them that is pulling them together. Eventually something's got to give.

In addition, we have a speed limit involved here: THE SPEED OF LIGHT

SPEED LIMIT 186,000 miles per second Strictly Enforced!

The important question is what happens in a situation where it's impossible to follow all the rules at once. Which rule predominates? We believe it will be the speed of light. When greater velocities are required among the particles in order for them to obey the Exclusion Principle, they will stop obeying the Exclusion Principle. That leave gravity the victor.

Figure 4. Games particles play: the exclusion principle

What is spacetime?

We use the term *spacetime* because the movement of any object is really in four dimensions, not in only three. We think of ourselves as moving through space in three dimensions. But even when we're perfectly still, we're always moving in time, the fourth dimension.

To apply this in a simple way: When we want to pinpoint our position exactly, we have to say not just where we are in terms of the three dimensions of space, but what time we are there. For instance, if I contact you by radio from a plane and report my latitude, longitude, and altitude, you'll know where I am in space. If I also tell you what time I'm there, that will be my position in time. You will know where I am in spacetime. This might be very useful, since my position in space is certainly changing rapidly as time passes.

Don't be concerned if you have trouble thinking in four dimensions. Most people do, although it's interesting to try.

THE AWESOME POWER OF GRAVITY

Why, sometimes I've believed as many as six impossible things before breakfast.
—*The White Queen in* Alice Through the Looking-Glass *by Lewis Carroll*

When gravity becomes enormously strong, peculiar things happen.

We're all familiar with the force of gravity on the earth. Let's pretend you take a trip into outer space. Figure 5 shows how things are on earth when you leave it. Let's further imagine that during your absence something odd happens to the earth—it gets squeezed. Now it's only half its original size, but it still has the same mass, pressed together more tightly.

As you return from your space journey and your rocket ship hovers for a while at the place, now in space, where the earth's surface used to be before the squeezing, the strength of gravity there is the same you were familiar with on the earth before you went

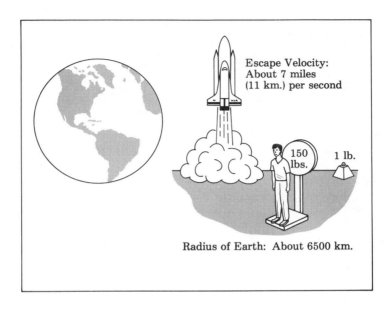

Escape Velocity:
About 7 miles
(11 km.) per second

150 lbs.

1 lb.

Radius of Earth: About 6500 km.

Figure 5. Earth's size as is

on your trip. But when you land on the new surface (a much smaller radius, quite a bit further in), the gravity there has quadrupled (Figure 6).

Let's say, instead of a half, that the earth is squeezed to a fourth its original diameter. Gravity on the new surface will then be sixteen times as great as you remember it. What if it were squeezed to the size of a pea? All the mass of the earth, billions and billions of tons, squeezed into that tiny space. Gravity on its surface would be so strong that **escape velocity** would be greater than the speed of light. Even light couldn't escape. The earth would have become a black hole. At the circumference out in space where the surface of

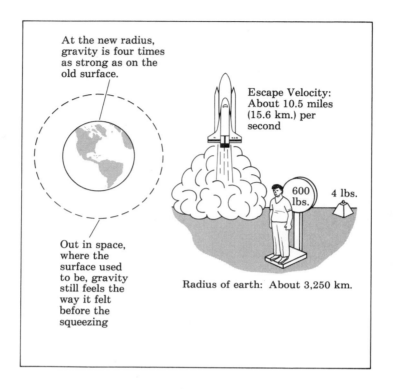

At the new radius, gravity is four times as strong as on the old surface.

Escape Velocity: About 10.5 miles (15.6 km.) per second

600 lbs.

4 lbs.

Out in space, where the surface used to be, gravity still feels the way it felt before the squeezing

Radius of earth: About 3,250 km.

Figure 6. Earth shrunk to half its present radius

the earth had been before the squeezing, the gravity would still feel just like it does to you and me today. This is an important point. You can see that black holes can't cruise around the cosmos like giant vacuum cleaners, sucking everything in. Keep a decent distance away and a black hole won't affect you any more than any other body with the same amount of mass (Figure 7).

Just as we've seen happening in this fantasy about the earth, the gravity on the surface of a collapsing

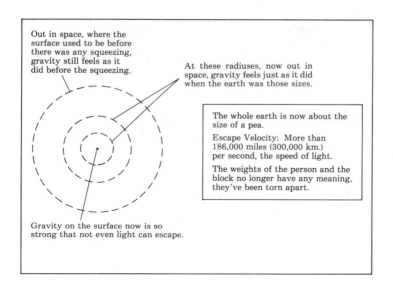

Out in space, where the surface used to be before there was any squeezing, gravity still feels as it did before the squeezing.

At these radiuses, now out in space, gravity feels just as it did when the earth was those sizes.

The whole earth is now about the size of a pea.

Escape Velocity: More than 186,000 miles (300,000 km.) per second, the speed of light.

The weights of the person and the block no longer have any meaning, they've been torn apart.

Gravity on the surface now is so strong that not even light can escape.

Figure 7. Earth shrunk to a black hole

star increases as the star shrinks. This has an effect on the paths that light follows when it travels nearby.

THE BENDING OF LIGHT

One way of thinking about light is to say that it's made up of particles called **photons.** Photons feel the pull of gravity. Whenever photons in a beam of light pass a massive body, such as a planet or a star, they react to its gravity. They move just a bit nearer to that body. We might say they are drawn slightly off course. If we trace the path they take through spacetime, we see that it bends in the direction of the massive body.

The massive body whose gravity bends the path of the light doesn't have to be anything as impressive as

31

a black hole or even a white dwarf. Take our sun, for instance. If a beam of light is traveling along a path from a distant star, and that path takes it close to the sun, the gravitational field of the sun pulls the photons in the beam a little nearer to the sun as they pass. The path the light is taking through spacetime bends inward toward the sun a bit.

Perhaps the path of light will bend in such a way that the light finally hits the earth. The sun is too bright for us to see such starlight except during an eclipse of the sun. But if we see it at that time, we might be fooled. If we don't realize the sun is bending the path of a star's light, we're going to get the wrong idea about which direction the beam of light is coming from and where that star actually is in the sky (Figure 8).

Imagine then what happens when the gravity increases on the surface of the massive body, in this case not the sun, but a collapsing star. The paths of light passing near it bend more and more. They begin to curl around it. Soon, they bend so far that they lead right into the collapsing star. Finally, any ray of light coming from the star itself is also drawn back in. No escape. Escape velocity from the star's surface has become greater than the speed of light. Nothing can exceed the speed of light, not even light itself. The star becomes invisible.

Another way to think about what's happening to the light is to say that its path looks bent to us because spacetime itself is "curved," or "warped."

Picture a new smooth foam mattress that's never been slept on (Figure 9). You could roll a marble straight across this mattress with no problem. If this is how the universe is designed, light will move in a way we recognize as a straight line.

Now suppose we drop heavy (massive) balls onto the foam mattress. They make dents and pockmarks

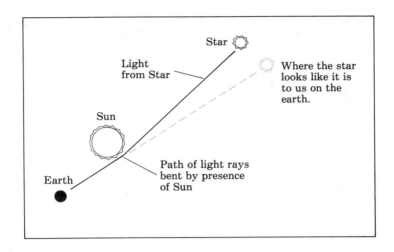

Figure 8. Light bending

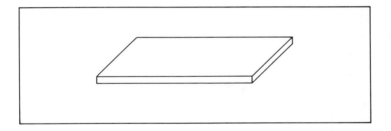

Figure 9. New mattress

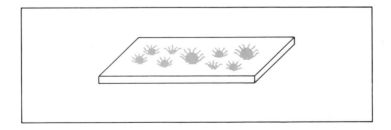

Figure 10. Dented mattress

where they lie. Some might be superheavy and tiny enough for the foam to close all the way over them, leaving only a deep dimple in the surface (Figure 10). You would have a tougher time rolling the marble straight across this mattress. Its path would be deflected and bend right or left every time it encountered one of the depressions caused by the balls. If this is how the universe is designed, there will be almost no such thing as a straight line as we know it. Anything going "straight ahead" (a beam of light, for example) will appear to be following a bent path. Planets will move in elliptical orbits and cause a slight dent of their own as they do so.

The second mattress is a rough model of the universe as Albert Einstein believed it really is. The balls represent massive bodies in space. Their mass causes dents and dimples in spacetime. They warp spacetime. Very massive ones that are also very tiny disappear entirely from the universe but still affect the warp. They are black holes, where spacetime curves out of sight. Planets do move in elliptical orbits.

If you're thinking that gravity causes the warping, that isn't exactly right. Warp of spacetime is a way of explaining gravity—not as a "force" but as something that happens because of the way spacetime is shaped by the presence of every single bit of matter in it.

It isn't easy to imagine light becoming invisible, in any sort of universe. Consider what we know about light and "seeing." You can think of light as particles (photons) or as electromagnetic waves. Either way of thinking is correct. Right now, think about it as waves.

What our eyes detect—when we say we are seeing—are waves of light. These are emitted from objects around us or bounced off them. To put it another way, we can't see anything except what arrives in our eyes by means of these waves. As a matter of fact, even *then* we can't see anything unless it comes in the colors of the rainbow or combinations of those colors. That's one of our limitations as human beings.

There are many other waves in the room with you—radio waves, for instance. You don't know radio waves are there until you turn on the radio, but they're there of course, nevertheless. Radio waves are electromagnetic waves (or photons) too, but not within our **visible spectrum.** That's how we're able to detect things out in space with radio telescopes that can't be seen with optical telescopes (Figure 11).

If a collapsing star sucks back all its light, no waves (or particles) come in the right direction to reach someone watching from the outside. The star becomes a black hole. There's absolutely nothing coming from it that our eyes are able to receive. The black hole swallows its own "picture" in space, the picture that would normally reach our eyes. The black hole similarly swallows radio signals, **X rays,** gamma rays. They're all electromagnetic waves/photons, and none can escape from a black hole. There will be nothing for a radio telescope or other receiver on the outside to receive.

THE CAPTURE OF LIGHT

Let's look at what happens at the moment the star swallows its picture in space. The mass of the star is getting compressed into a smaller and smaller space, to greater and greater density. Gravity on the star's shrinking surface is growing fiercer and fiercer. (We learned earlier how this would happen if the earth were squeezed.) The paths of light from other stars are bending in more and more toward the star. It's getting more and more difficult for light coming from

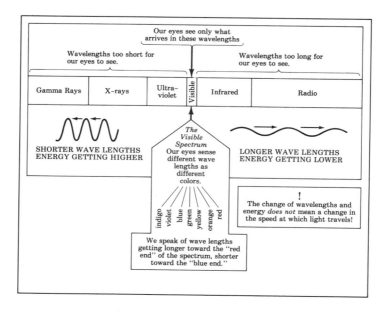

Figure 11. The electromagnetic spectrum

the star itself to escape into space. This light, the light coming from the star itself, is what we're interested in now.

There will be an instant during the collapse of the star when the **curvature of spacetime** at the star's surface is almost but not quite great enough to bend light coming from the star all the way back in. Photons (particles of light) travel at the speed of light, and the escape velocity needed to get away from the star's surface is not quite that great yet. The last photons that will ever escape from the star are escaping in this instant, making the break for freedom with the beast of gravity nipping at their heels!

36

Then, in the next instant, the surface will have shrunk a little farther in and spacetime curvature there *will* be great enough to end all escape. Escape velocity now is greater than the speed of light. Any photon coming out of the star will be pulled back.

There is a very important split second in between these two instants (between the instant-of-last-escape and the instant-of-having-to-get-pulled-back-in). In this split second, the collapsing star officially becomes a black hole. Spacetime curvature on the surface of the collapsing star becomes for that split second just enough to keep photons from escaping, but not strong enough to bend their paths all the way back in. The photons emitted in this split second are caught right there and mark the boundary of the black hole. They will have to stay at that radius as the star goes on shrinking. They won't get away (as did those that came before the surface had shrunk quite so far). They won't be pulled back in (as do those that come when the surface has shrunk further). These photons swarm in a thin, spherical shell, like a membrane surrounding the interior of the black hole. The star goes on shrinking inside this shell.

Because this is a little difficult to understand, here's the same thing put slightly differently: Remember when the imaginary earth got squeezed? At the place in space where the surface was before the squeezing, the strength of gravity stayed exactly the same, even when the earth got to be a black hole. The strength of gravity at the radius where these photons are hovering is just exactly the right strength to keep them from getting away and at the same time to not pull them in. If the photons were the tiniest bit closer in, they would be pulled in. If they were the tiniest bit farther out, they could escape. Where they are, they can only spin their wheels.

Don't let this confuse you. The gravity *on the surface* of the collapsing star is still growing stronger. But if you take any distance out in space from the star's center, the gravity at that **radius** is going to remain what it was when the star's surface was at that radius.

Unless the mass of the shrinking star changes, this spherical shell of hovering photons will always stay just where it was when it first formed. Regardless of how much the star shrinks *inside it*, the gravity at just this radius, where these photons are, will not change.

However, it's possible that more matter *will* fall into the black hole. More matter means greater mass, and that means a stronger gravitational pull. "All the way out," gravity or spacetime curvature will increase over what it was before. The original shell of hovering photons will get pulled in. There will be a new, slightly larger, radius at which escape velocity exceeds the speed of light. The black hole will have got slightly fatter.

Can it ever get *less* fat? Can it lose mass? A little hard to do, if nothing can escape! It's pretty safe to say that, except in rather special circumstances, which we'll encounter in Chapter Four, a black hole isn't going to get smaller.

Meanwhile, we must give this sphere or membrane where the shell of hovering photons is, its correct name. It's called the **event horizon.** The horizon on earth is the line in the distance caused by the curvature of the earth. You can't see what's going on beyond it. At the horizon, the earth's surface curves out of sight. The event horizon of a black hole is caused by the curvature of spacetime. You can see the events that take place on our side of it, but unless you cross it, you will never see the events that take place on the other side of it. At the event horizon, spacetime curves out of sight.

If light can't outrun the gravity of a black hole, can anything else? A spaceship, for instance? For a spaceship getting away from the earth's gravity, escape velocity (the speed needed to get away) is about 7 miles (11 kilometers) per second. For anything trying to get away from the event horizon of a black hole, escape velocity is a little bit more than the speed of light. (You'll recall that photons, traveling at the speed of light, are held captive there.) According to Einstein nothing moves faster than light. So nothing can outrun the gravity of a black hole. If you cross the event horizon, you'll never return. You'll never be able to tell the rest of us what it's like in there. Nothing whatsoever can escape from inside the event horizon to reach us on the outside.

At the exact center of a black hole, inside the event horizon, all the matter of the star is finally compressed to a point that has no dimensions at all. A point of infinite density . . . a **singularity.** The paths of light at the singularity are wound infinitely tight: the curvature of spacetime is infinite. The laws of our physics break down. We reach the end of space and time as we understand them (Figure 12).

Now we're in deep water: singularity, infinite density, infinite curvature of spacetime, end of space and time. Our minds balk at such notions. The most highly regarded physicists in the world are not far different from you and me at this point. We're all out of our depth.

We should review the terms "event horizon" and "singularity" because we'll be using them often as we go on.

The event horizon is the invisible outer membrane or shell of the black hole. Think of it as the outside edge of a spherical (ball-shaped) part of space with the singularity dead center within it. It isn't possible for anything that can't go faster than the speed of light

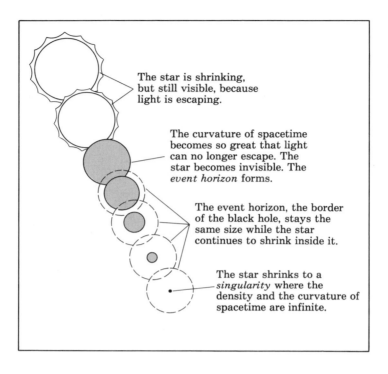

The star is shrinking, but still visible, because light is escaping.

The curvature of spacetime becomes so great that light can no longer escape. The star becomes invisible. The *event horizon* forms.

The event horizon, the border of the black hole, stays the same size while the star continues to shrink inside it.

The star shrinks to a *singularity* where the density and the curvature of spacetime are infinite.

Figure 12. A star collapses and becomes a black hole

to escape from inside the event horizon: no rocket ship, no astronaut, no radio signal—nothing at all. Photons at the event horizon can't be pulled in and can't get away. They hover there forever (Figure 13).

A singularity is the dimensionless point at the exact center of the black hole. Here, all the mass of the collapsing star has shrunk to infinite density. The curvature of spacetime is infinite. Anything falling into the black hole will be drawn to the singularity. When it gets there, it will have reached the end of space and time as we understand them.

What's it like in there, in the area between the

How can photons be "caught and held" at the event horizon without the speed of light slowing down.? The "increasing rapids" represent the increasing curvature of spacetime. if the boats are photons, and Singularity Point is the singularity at the center of a black hole, it's easy to see which photon is outside the event horizon and getting away into space—P1. P3 is inside the event horizon and being pulled back to the singularity. P2 is going to hover forever at the event horizon.

All photons maintain a "rowing speed" of about 186,000 miles (300,000 kilometers) per second. (They are very good athletes!) But inside a black hole, the "current," whch we call gravity or the curvature of spacetime, is too much even for them. At the event horizon, rowing for dear life, they CAN just barely hold their own.

There are three rowers, all required to
row at the same speed—20 mph.

TO
SINGULARITY POINT
BEWARE INCREASING
RAPIDS!

The boats move
upstream, away
from Singularity
Point, against the
current.

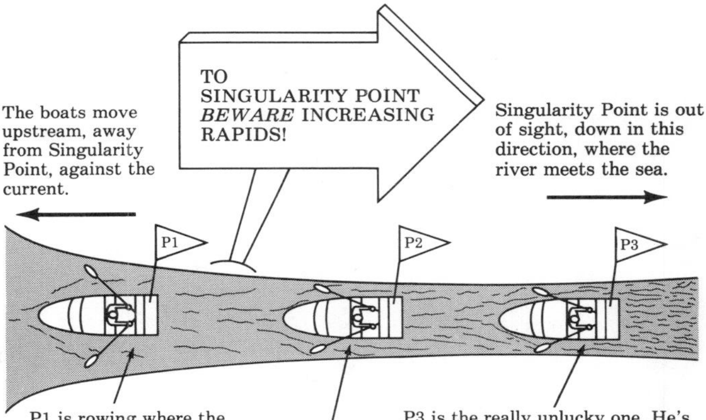

Singularity Point is out
of sight, down in this
direction, where the
river meets the sea.

P1

P2

P3

P1 is rowing where the
current isn't strong.
He's maintaining a row-
ing speed of exactly 20
mph and he's moving
upriver.

P3 is the really unlucky one. He's
rowing where the current is very
very strong. He's maintaining a
rowing speed of 20 mph, but the
current is pulling him back downriver toward Singularity point.

P2 is in a frustrating position. He also is maintaining his
rowing speed of 20 mph, but when he glances to the right
and left, he sees that he's staying in precisely the same
spot in relation to rocks and trees on the bank. The
current where he's rowing is just strong enough to hold
him where he is as long as he rows at 20 mph. It isn't
powerful enough to pull him back downriver toward the
sea and Singularity Point. It isn't weak enough for him
to make headway upriver. He's stuck there unless the
current changes.

*Figure 13. Games particles play:
the regatta near the event horizon*

event horizon and the singularity? What actually happens at the singularity? Wouldn't we all like to know!

Just *outside* the event horizon things are relatively normal. The laws of our physics are stretched almost to the breaking point, but they still hold. That means we can make predictions about what it would be like to go there for a visit.

3
TREACHEROUS VOYAGE

When I examine myself and my methods of thought, I come close to the conclusion that the gift of fantasy has meant more to me than my talent for absorbing positive knowledge.

—*Albert Einstein*

When someone tacks up a recruiting poster for the first expedition to a black hole, don't be too quick about signing on. "Exploring strange new worlds where no one has gone before" sounds great. Exploring near an area from which no one can return might be less inviting (Figure 14).

Such a journey may never take place. It can't take place with our present technology. However, fantastic voyages of exploration are possible in our minds. They needn't be science fiction. There's quite enough to boggle the imagination within the bounds of well-established theory. This sort of speculation is part of what theoretical physicists do for a living.

43

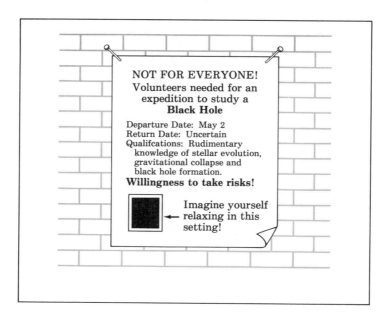

Figure 14.

Let's try to visualize for ourselves what a journey to a black hole might be like, based on what those who study black holes know or think they know about them right now.

Some ground rules:

1. Don't ask how we get there or, when we do, how we survive the turbulence near a collapsing star. We'll leave the design of the ship to the technology of the future.

2. Don't ask how we can possibly know with pinpoint accuracy that a particular star is going to collapse at a particular instant, and that it's going to do it without a supernova which no spacecraft could survive. Trust our descendants with that, too.

3. Keep in mind that it's always risky to take theory and translate it into "what we would see" or "what we would feel." There are too many ifs, ands, and buts about it. We'll do our best.

Agree with me that somehow or other we've gotten ourselves into orbit around a star that's on the brink of gravitational collapse. We'll say it's a massive star, about one hundred times the mass of our sun back home.

Note that the mass is well above the Chandrasekhar limit or the limit for neutron stars. With any luck, we are about to witness the birth of a black hole.

The orbit of a spacecraft around such a collapsing star must be calculated with great precision. There is considerable danger, some of it unpredictable. What do we know about such hazards?

TIDAL EFFECTS

The ocean tides make all of us familiar with **tidal effects** on earth. The gravity of the moon pulls on the earth more on the side closest to the moon. This pull stretches the earth out of shape just a bit—first this way and then that, depending upon where the moon is in its orbit. We don't notice that stretch except where there are great bodies of water. Whatever part of the ocean is closest to the moon bulges out toward the moon, enough to make a high tide. Whatever part of the ocean is farthest from the moon and feels the least pull gets left behind a little and bulges out *away* from the moon—another high tide. Halfway in between we get low tide (Figure 15).

Tidal effects are relatively harmless in that form. Not so in the vicinity of a black hole.

Near a black hole, gravity becomes enormously strong. This in itself won't harm us. In orbit, in **free**

The "shell" of ocean around the Earth gets stretched out of shape by the pull of the moon's gravity. Point "A" feels this gravity more strongly than any other part of the earth. Point "B" feels it least, and gets left behind. We get high tides on both sides of the earth. It's the *difference* in the pulls, not the direct gravity itself, that raises the tides.

Figure 15.

fall, we'll feel weightless. However, the strength of gravity increases rapidly as you approach the event horizon, the outer border of the black hole. It may increase so rapidly and over so short a distance that the force of gravity on one side of the spacecraft, the side facing the black hole, will be vastly greater than on the other side.

The closer a spacecraft, or anything else, approaches the hole, the greater this difference in pulls on two parts of its structure will be. Ultimately the spacecraft won't be just stretched like the earth and its oceans; the spacecraft, and anyone inside it, will be torn apart. The effect will be felt most drastically at first by the largest objects (their two sides are farther apart than those of smaller objects). But near the singularity, the point at the center of the black hole, the very particles that make up matter will not be able to hold together.

A safe orbit for a spacecraft must take into account

the ability of the ship and its passengers to tolerate various degrees of stretching. It must also take into account the size of the black hole. If it's small, the event horizon is close to the singularity, and tidal effects are enormous even outside the event horizon. If it's large, like mammoth black holes in the centers of galaxies, the event horizon is very far from the singularity, and survival (but not escape) might be possible well within the event horizon.

The star we're imagining (100 solar masses), doesn't fall into the second category. If we come too close, tidal effects will make for ugly possibilities.

HAZARDS OF ROTATION

It's likely that our star is rotating. If it is, the rate of rotation after it's collapsed to a black hole may turn out to be at least half the speed of light. Does such a rate of rotation seem preposterous to you? A star's rotation isn't usually all that astonishing. Why should it speed up so much when it becomes a black hole?

Think about a skater doing a spin. She starts the spin on one skate, with the other leg and both arms extended. Then she pulls her leg and arms tight in against her body. When the mass of a spinning object draws closer to the center of the spin, the **axis,** the speed of the spin gets faster. As the skater draws in her leg and arms, she spins faster.

We can demonstrate the same thing on a playground merry-go-round. When everybody sits on the outer edge and leans far out away from the center, the merry-go-round almost comes to a stop. If everybody moves toward the center of the merry-go-round, it speeds up. It doesn't seem possible, but it's no illusion. It seems like getting something for nothing, but that's not the case. Instead, it's hanging on to something that's already there—the **angular momentum.**

We say that angular momentum is "conserved," or saved. What this means is that no angular momentum can simply appear out of nowhere or evaporate and disappear. When it isn't needed to pull around parts of a body that are far out from the axis, it has to be used in another way. It goes into speeding up the rate of rotation.

The collapse of a star to a black hole is a mammoth drawing in of arms and legs. The black hole will still have the angular momentum that the star had, but the mass will be concentrated *very* near the axis. If the star is spinning at all, chances are the black hole will have a stupendous rate of rotation.

A rotating body affects things in spacetime around it—such as spaceships, gases, particles—in an odd way. You can demonstrate something like this by spinning a ball in water (although spacetime definitely isn't a substance like water). Watch how the spin of the ball affects anything floating near it. The faster the ball rotates, the stronger the effect.

Objects near a rotating black hole feel this effect in a powerful way, increasingly powerful the nearer they are to the event horizon. Somewhere still outside the event horizon, the dragging becomes so overwhelming that nothing can possibly remain at rest. Anything there is swept around in the same direction the hole rotates. Blasting rocket engines as hard as you can won't do you a bit of good; you won't be able to resist this dragging. The area around a rotating black hole where the dragging becomes irresistible is called the **ergosphere.** Its outer edge is the **static, or stationary, limit,** the limit inside of which you can't remain stationary (Figure 16).

How will this affect our ship? Outside the static limit, we can maintain a position (not an orbit, a position) that allows us always to see the same pattern of

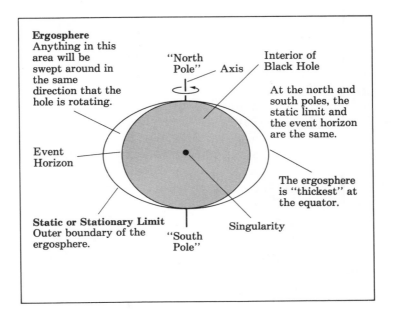

Ergosphere
Anything in this area will be swept around in the same direction that the hole is rotating.

"North Pole" Axis

Interior of Black Hole

At the north and south poles, the static limit and the event horizon are the same.

Event Horizon

The ergosphere is "thickest" at the equator.

Static or Stationary Limit
Outer boundary of the ergosphere.

"South Pole"

Singularity

Figure 16. Ergosphere

stars while the black hole rotates beneath us. All that's needed to maintain this position is a powerful enough downward blast of our engines. Between the static limit and the event horizon, within the ergosphere, such a position isn't possible. Our ship would be dragged around in the direction of the rotation of the hole. We could manage to go faster or slower, but go we would.

HAZARDS OF GRAVITY

Gravity, like light, can be thought of either as waves or as particles, called **gravitons.** When a star collapses to a black hole, theory predicts that there will be a short burst of gravity waves. This ripple or wobble in the fabric of spacetime will be felt by those on board our orbiting ship as a short period of intensified tidal effects. If the collapse is very neat and near-spherical, this wobble won't be so great. But realistically, we have to expect that rotation and turbulence in the collapsing star may cause an impressive and potentially dangerous burst of gravity waves.

Passengers can be grateful that the burst of gravity waves is so short. Why won't it continue for a longer time? The answer to that can tell us some interesting things about black holes.

First, the collapse happens quickly, the last stages within a microsecond. Blink, and you'll miss it. Second, **gravity waves** can't escape from inside the black hole after the star has shrunk within the event horizon. There's another curious factor at work, a factor which becomes particularly important around black holes and other areas of extreme gravity.

GRAVITATIONAL REDSHIFT

You've probably learned that if something moves toward us, sound waves coming to us from it get bunched together. If something moves away from us, sound waves coming to us from it get stretched out— the distances between the wave crests get longer. Longer sound waves make for lower pitches. A siren drops to a lower note when it has passed us. We call this familiar phenomenon the **Doppler effect.**

The Doppler effect also works with light waves.

The longer the **wavelengths,** the further we move to the red end of the electromagnetic spectrum. It's called a **redshift.** If something continues to move away from us, *increasing its speed as it goes,* the waves eventually get stretched to such lengths that they move beyond the visible spectrum. The object becomes invisible. (In everyday circumstances, we don't notice a shift to the red end of the spectrum as objects move away from us; only at speeds nearing the speed of light would it become noticeable to our eyes.)

The Doppler effect also works with gravity waves—they get stretched as the source of them accelerates away from us.

Will the collapsing star be moving away from our spacecraft? No, not the star. But the *surface* of the star nearest to us will be shrinking away from us more and more speedily as the star collapses. Think of a balloon. If you let the air out while you hold on to the balloon, the balloon won't move away from you. But the *surface* of the balloon nearest to you will move away.

There's another way besides **acceleration** to stretch sound or light or gravity waves. Einstein made the remarkable observation that *gravity acts like acceleration*. The two feel the same. They have the same effect on things.

Here's a familiar example: Think about what it feels like to ride an elevator from the first to the thirty-fifth floor. The elevator door closes. For a few seconds you feel heavier. You feel as though gravity is pulling down harder on you than normally. You're being fooled. You feel heavier because the elevator is accelerating—going faster and faster—as it rises. Then the elevator reaches its maximum speed and the speed becomes constant. No more acceleration. You feel only the normal pull of the earth's gravity, holding you comfortably to the floor of the elevator. When the elevator nears the thirty-fifth floor, it begins to slow down and you experience deceleration, or negative acceleration. You feel as though gravity has gotten weaker. You feel lighter.

You know very well that during the time you spent

51

riding the elevator the gravity of the earth didn't change, but because of the effects of acceleration and deceleration on your body you felt as though it had (Figure 17).

Another example: In a spacecraft out in space far away from any planet or other astronomical body, traveling at a constant speed in straight-line motion, the passengers float gravity-free. In a spacecraft that continually accelerates, the passengers feel a pull like gravity. They may feel heavier or lighter than they would on earth, depending upon the rate of acceleration. Be that as it may, one surface of the passenger compartment, the one in the opposite direction from the direction of the acceleration, will be "the floor." Everything on board—liquids, rubber balls, photons—will act as though that were indeed the floor. If the rate of acceleration is just right to simulate the earth's gravity, and if there are no windows and no engine noises or vibrations, the passengers won't know whether they're in outer space or still on the launch pad (Figure 18).

Gravity and acceleration have a near-identical effect on people, on light waves, and, we think, on gravity waves. Gravity stretches waves as surely as acceleration does. We call the result a **gravitational redshift.**

If gravity acts on gravity, what will happen to a gravity wave trying to escape a collapsing star as the gravity of that star gets more and more powerful? Does that sound like double talk? Well, it almost is. It's much easier to think about light waves coming from the collapsing star:

As the star collapses, we'll see it look redder and redder. Why? Increasing gravity on the star's surface and the surface's acceleration away from us stretch the waves toward the red end of the spectrum. Even

52

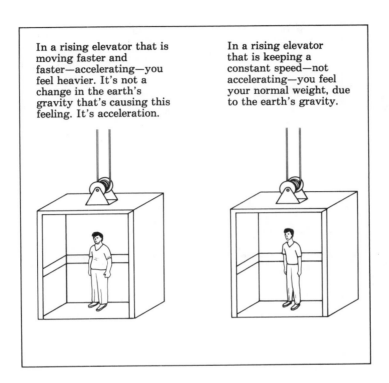

In a rising elevator that is moving faster and faster—accelerating—you feel heavier. It's not a change in the earth's gravity that's causing this feeling. It's acceleration.

In a rising elevator that is keeping a constant speed—not accelerating—you feel your normal weight, due to the earth's gravity.

Figure 17.

before the star collapses enough to form the event horizon, the light will have redshifted out of the visible spectrum and even beyond the radio spectrum.

If you thought you would see the star "wink out" as it formed the event horizon, you were wrong. It will become invisible before that. The wavelengths will get too long for us to detect with *any* equipment. At the event horizon the redshift becomes infinite.

Another way of thinking about this is that light uses up all its energy getting away from the edge of the black hole. You can see from the diagram on the

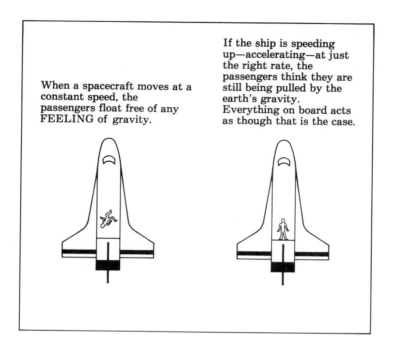

When a spacecraft moves at a constant speed, the passengers float free of any FEELING of gravity.

If the ship is speeding up—accelerating—at just the right rate, the passengers think they are still being pulled by the earth's gravity. Everything on board acts as though that is the case.

Figure 18.

spectrum (Figure 11) that lower energy levels go along with movement toward the red end of the spectrum. Photons from near the event horizon reach us, arriving at the speed of light but with no energy left.

That's how it works with light. Now back to the burst of gravity waves. Here it comes, as the star collapses. But then it weakens, its waves redshifted more and more, until we can't detect it at all. After the event horizon forms, no more **gravitational radiation** will come from inside the black hole.

In spite of the fact that no gravitational radiation will be escaping from within the event horizon, the mass of the black hole will still have a gravitational

54

effect on other objects such as our ship or another star. It will continue to warp spacetime, a dimple in the foam mattress.

What we really want to know is how dangerous it will be when the burst of gravity waves hits our ship. The redshift will save us, *if* our orbit isn't too close. The collapsing star will snatch back its own gravity just in time to rescue us from tidal effects that we couldn't survive.

As a precaution, as the star collapses, we ought to position our ship, and perhaps our bodies as well, so that the part nearest the black hole is the least possible distance from the part farthest from the black hole. If you recall the discussion of tidal effects, it isn't the pull of gravity that's a problem. In fact, we feel weightless in our free fall orbit. The danger comes from the *difference* in that pull on two parts of the same body. The closer the two parts, the less drastic the stretch. If you're standing upright, with your feet toward the black hole, the pull of gravity on your feet might be much stronger than the pull on your head. It might feel like being stretched on the rack in a medieval torture chamber. It could be fatal.

PREVIEW OF THE COLLAPSE

From now on, we'll occasionally be using diagrams with a **time-line**. If such a thing is unfamiliar to you, you should study the first one below with special care. It shows the final collapse of a star (Figure 19).

Notice the arrow on the left side labeled "time." That's the time-line. It serves like a clock for the drawing. Start at the bottom of the diagram. The center area is the star. Follow the time-line up. You can see that the center area is getting smaller. The star is collapsing.

55

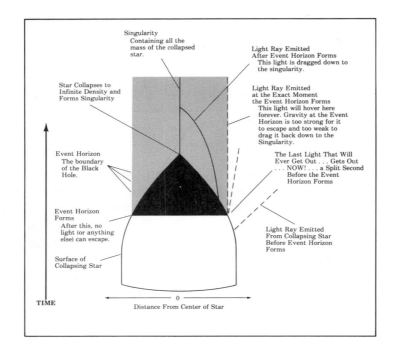

Figure 19. Final collapse of the star to form a singularity and a black hole

On the right-hand side of the drawing you see a light ray from the star heading off into space, perhaps to reach an orbiting ship. A little further up the time-line, the drawing shows that the event horizon has formed. We can see a light ray staying right at the event horizon—one of the light rays that *form* the event horizon. That one will never reach an orbiting ship or any other outside observer, nor will it be drawn into the singularity. Time marches on (up), and the star collapses to a point—a singularity.

You can also learn from the diagram that a light

ray emitted from the star after the event horizon forms doesn't get away into space. It's pulled to the singularity. At the bottom of the drawing, where the time-line begins, we still have a star. Before the time-line reaches the top of the drawing, we have a black hole. The transformation, as you will see, happens so quickly and at such a rate of acceleration that it will be necessary to photograph it and play the film back in slow motion in order to study it.

VOLUNTEER FOR OBLIVION

Whenever scientists or science writers write about black holes, they send an unfortunate imaginary astronaut down into an imaginary black hole. It could just as well be an unmanned probe, but somehow, in book after book, it's an astronaut. Human curiosity being what it is, it's probably fair to assume that someone would volunteer for that mission, if it were available, even though he or she would never be able to return and enjoy the publicity.

Let's not be the exception. Let's send down an astronaut from our ship, land him on the surface of the star, and have him stay there as it collapses to a black hole. This will, as you know, be a one-way trip. The plan is that after he reaches the surface, a microscopic radio built into his suit will send a beep every second.

THE COLLAPSE: FROM OUR POINT OF VIEW

In the real-life collapse of a star, our view of it would be obscured by dust and debris. However, for the sake of learning about black holes, we must imagine that we can somehow see what's happening:

We're on the bridge of the ship, watching the star on the screen and listening for the beeps from the astronaut's radio. The star is being magnified somewhat by the ship's instruments, because from our distance—a safe distance, we think—the last stages of the collapse won't be easy to see.

The ship shudders, and we sense a change in the deep rumble of the engines. Engineering is repositioning us to take the least possible damage from the increased tidal effects.

Three fifty-nine by our ship's clock. The beeps begin. The astronaut (or at least the radio) has reached the surface of the star. We count ten beeps. It's fifty seconds before four. We listen to the regular beeps, and on the viewing screen, sure enough, the star begins to shrink into itself faster and faster like water going down a drain. An ominous groan from the ship . . . a snapping sensation in our bodies. The shrinking stops, the star seems frozen in space. It turns red. Surely the fifty-ninth beep should have come by now. There it is, noticeably late. The star darkens and in a fraction of a second becomes only a shadow and then completely black. *Nothing.* Our ears expect the sixtieth beep, but there is silence. We will never hear the four-o'clock beep.

THE COLLAPSE: FROM THE UNFORTUNATE ASTRONAUT'S POINT OF VIEW

The astronaut lands on the star and switches on his radio. The beeps begin. The star begins to shrink. How does he know? Does it feel like an earthquake? At some point, maybe so, when matter becomes compressed to the density of a solid. He notices the redshift of distant stars as he accelerates away from them. Perhaps he notices a blueshift of stars behind him. The fifty-ninth beep doesn't seem late to him. The beeps sound regularly every second. The star shrinks below the event horizon, but there is no way the astronaut can tell this has happened. He and his radio fall to the singularity in a fraction of a second.

You're probably asking many questions now. Why did the radio beeps sound farther and farther apart to our ears but not to the astronaut's? Why did the beeps stop for us but not for the astronaut? Why did the star appear to us to freeze in space for a split second before disappearing? What has actually happened to the astronaut?

First, the question of the beeps. In order to understand this, I must warn you, you must put yourself in a frame of mind somewhat like the White Queen in *Alice Through the Looking Glass,* who sometimes "believed as many as six impossible things before breakfast."

WHAT'S WRONG WITH OUR CLOCKS?

We learn from Einstein's theories that time doesn't pass at the same rate everywhere in the universe.

No matter where you or I happen to be in the universe, time right there will seem to us to pass in its familiar manner. There is no place where the inhabitants see their time passing as it does in fast-forward on a video. But if inhabitants in certain parts of the universe could watch events in certain other parts, they would notice mind-boggling differences in the rate at which time passes.

Because of this bizarre effect, which we call **time dilation,** we can make two sets of statements, which disagree completely with one another but are nevertheless both true:

1. "The surface of the star and the astronaut and his radio take an infinite amount of time to

cross the event horizon. Time slows to a stop there. The surface of the star and the astronaut and his radio are stuck there forever. They never reach a singularity." These statements are true from the point of view of observers like us everywhere else in the universe at a distance from the black hole. From our vantage point, the slowing of time at the event horizon becomes infinite, which means that by our clocks out here, time there does stop.

2. "The astronaut and the star will cross the event horizon and reach a singularity in a split second. The singularity is a finite point, more a point in time than a point in space. They must reach it just as surely as we all must reach a point in time that we call, for instance, tonight at 11 P.M. It's unavoidable." These statements are true from the point of view of the astronaut himself. He'll cross the event horizon and get to the singularity very quickly indeed.

You can see that there's a shocking discrepancy here. Neither point of view—ours or the astronaut's—is "right" or "wrong." It all depends on where you are. We believe this is fact, not science fiction, but there's nothing in fantasy quite so incredible.

TIME DILATION AND REDSHIFT

Time dilation isn't easy to understand. It helps if you know that time dilation and redshift are two faces of the same coin.

You already know that acceleration and gravity can have almost identical effects on things. Both, we've seen, can stretch electromagnetic waves. For purposes of this discussion, remember that that includes light waves and radio waves. We've watched light waves coming from the collapsing star as they were stretched to longer lengths. The light moved toward the red end of the spectrum; it was redshifted. (See Figure 11.)

Since whatever we perceive about events at a distance comes to us by means of some sort of waves, redshifting of those waves means we perceive those events at a slowed-down rate. In other words, we have time dilation.

Let's break that down a bit:

Saying that wavelengths are stretched is the same as saying that the crests of the waves reach us less frequently. We're accustomed to hearing the term "frequency" used when we talk about radio waves. It may or may not have occurred to you that "frequency" means how "frequently" the wave crests arrive. If they arrive more frequently (shorter wavelengths), that means they have a higher frequency. If they arrive less frequently (longer wavelengths), they have a lower frequency.

Redshift, caused by acceleration or gravity, means that the wave crests are reaching us less frequently. The *frequency* of light waves, and the *frequency* of any events we perceive by means of these light waves, is lower.

To get back to our poor astronaut. He accelerated away from us with the surface of the collapsing star (see p. 57). As the collapse progressed, this acceleration became greater and greater, causing an increasing redshift and a slowing-down of time, by our measurement on the ship.

We perceived this slowing of time as greater distances between the radio beeps. The beeps traveled to us by means of waves in the radio part of the spectrum. Their radio frequency was slowed down by acceleration as the star surface moved more and more rapidly away from us. Because there was no way we could know about the beeps other than what the light waves told us, we perceived that they were taking longer to happen.

The growing distance between the beeps was small at first. But the speed of the collapse kept on increasing. Between the 3:59:58 beep and the 3:59:59 beep, the delay was great enough for us to notice. The interval between the 3:59:59 beep and the 4:00 beep is forever—for us, where we are. From the vantage-point of our ship, time at the event horizon never reaches 4:00. Time there has come to a complete stop.

Why didn't the astronaut see that this was happening? Remember that none of this "message" was slowed down right where he was, where it started out. If you're at the bottom of a deep pit, the steepness of its side doesn't slow you down until you try to climb out. Light waves, in the same way, aren't slowed down until they start to climb away from the black hole, out of the "pit." The astronaut (down in the pit, so to speak) didn't see them slowed down. For him, time didn't seem as though it slowed down at all. He had no way of finding out that during the split second it took him to cross the event horizon, your watch and mine show that billions and billions of years pass by out in the area where our spacecraft is (or was) orbiting.

All of this sounds as though time dilation isn't "real." Isn't it only an illusion, a trick played on our eyes and ears? No, it is real.

There's a small gravitational redshift and time dilation in the short distance from the bottom to the top of the Empire State Building.

Try this: You stand on 34th Street, New York City, at the tourist entrance to the Empire State Building.

I'll stand on the observation deck at the top. After several hours, we'll meet and compare watches. (These will have to be *much* more accurate watches than anything we own at present.) Yours will be a tiny fraction of a second behind mine. This happens because of the small gravitational redshift. The gravity of the earth is felt more strongly on the ground, where your watch is, than at the top of a tall building, where my watch is. (Remember, gravitational attraction grows weaker with distance.)

The difference in the speed of clocks at different heights from the earth has practical importance. We have superaccurate navigation systems based on signals from satellites. If we ignored time dilation, our calculation of position could be off by several miles. Time dilation is not an illusion!

Imagine what happens to time near the event horizon of our new black hole, with all its enormous gravitation. When we on the spacecraft measure it by our watches, we see time nearer the black hole terribly slowed down compared with where we are. In fact, at the event horizon, we see that it stops. On the other hand, if we were able to sit just outside the event horizon and keep in touch with events back on earth, we would see those events moving forward at a pace that would amount to a blur.

FROZEN IN TIME AND SPACE?
THE COLLAPSE IN SLOW MOTION

Why did the star seem to freeze before it disappeared?

You might think we've just answered that question: Didn't time dilation slow down our picture of the collapse (a picture arriving at our eyes by means of light waves) until we couldn't tell it was moving at all?

It's true that that happened, but by then the light

was beyond our visible spectrum. We didn't see it happen. We must look for another explanation for the "freezing."

Pretend we've taken the video of the collapse to a viewing room to watch it in slow motion—not the slow motion caused by time dilation, slow motion caused by the "slow" button on our video machine.

What will we see? Because rotation makes the following picture much more complicated, we'll say the star was *not* rotating.

First we see a bright disk getting smaller but not dimmer or redder. The change in size is slow at first, then faster and faster. Then the disk stops getting smaller. At its center we see a dark red spot that gets darker until it's black. This spot eats its way out into the bright rim. The rim gets thinner and thinner and finally becomes invisible (Figure 20).

What happened to cause this odd picture?

Remember that not every photon coming from the star is coming straight out—radially—like the spokes of a simple wagon wheel (Figure 21). Instead, like most things that radiate light, the star radiates it in many, many directions at the same time. Most photons come out at an angle, like some spokes on the wheel of a fancy sports car (Figure 22). Photons coming out at angles like that have less chance of getting away than those coming straight out. Imagine the boat race again. If a boat joins the race at an angle, not pitting all its speed directly against the current, its chances of success are reduced (Figure 23).

Think of the photons coming out at an angle (the sports car types). Spacetime curvature bends the paths of some of them all the way back in, even before we reach the event horizon stage. Other paths aren't bent quite so far as that, but bent far enough to get

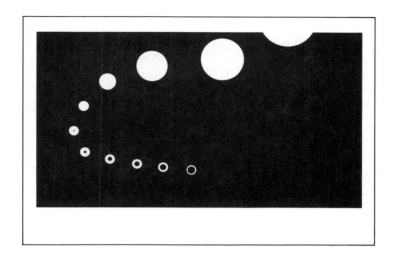

Figure 20. Collapse in slow motion

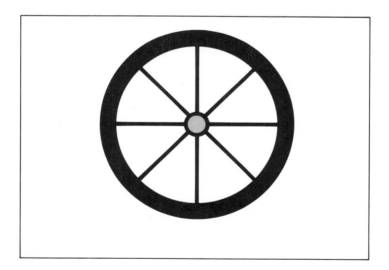

Figure 21. Spokes of wagon wheel

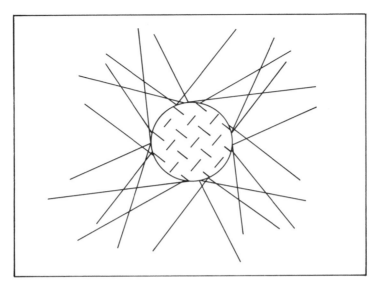

Figure 22. Spokes of sports car wheel

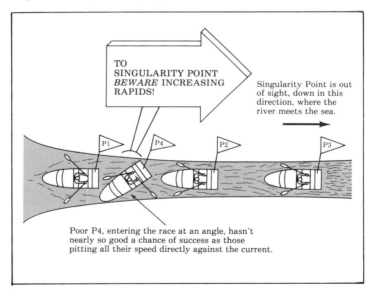

Figure 23. Boats in river with one turned sideways

captured for a while in a circular orbit around the collapsing star. A circular orbit isn't stable, and those photons leak out to our eyes gradually. As they're doing that, we see them as the bright rim.

Meanwhile, those photons coming straight out at us radially (the wagon-wheel types), aren't captured in the circular orbit and reach us more quickly. We can't see many of these photons from any one position out in space, of course. (When you look at the edge of a wagon wheel, most of the spokes aren't pointing at you.) Soon there aren't any of them left except some that are too redshifted to see. There are no more photons in the visible spectrum coming straight out at us. No more photons in the visible spectrum means blackness, and we see the dark dot in the center of the disk. Eventually all the photons in the circular orbit finally leak out to us too; then we see nothing at all. During the collapse most of this happened in less than a microsecond.

A WORMHOLE FOR OUR ASTRONAUT?

When we discussed time dilation, and how time would pass differently for us on the ship and for the astronaut entering the black hole, we ignored the fact that the astronaut was unlikely to be able to appreciate any of this. Gravity gets stronger the closer you are to the singularity. Think of the tidal effect. He was surely strung out like spaghetti and torn apart even before the event horizon formed.

Whatever's left of him will almost certainly find itself pulled down to the singularity—the point where there is infinite density, the point where spacetime curvature and the bending of the paths of light are also infinite. The end of spacetime as we conceive of it.

Could our astronaut escape through something called a **wormhole?**

It may be possible, if the black hole is rotating, to miss the singularity, slip through a little hole in space-time, and end up in another part of the universe, in another universe, or somewhere else in time.

Great science fiction!

Unfortunately, wormholes aren't a reliable way to travel, even if the astronaut could get to one without being shredded by tidal effects, which he can't. Any disturbance, such as the presence of an astronaut, will destroy a wormhole.

Physicist Stephen Hawking says, on the other hand, that *particles* may travel regularly through wormholes. Friends of our astronaut may console themselves by hoping that the particles which made up his body might be recycled in another universe.

4
CONTEMPLATING AN ENORMOUS NOTHING

A huge great enormous thing, like—like nothing. A huge big—well, like a—I don't know—like an enormous big nothing. . . .
—Piglet describes the Heffalump,
in Winnie the Pooh, *by A. A. Milne*

At this point, we on our imaginary ship are no longer looking at a star, not even a collapsing star. We're looking at a black hole. Our science staff will get busy studying it, measuring it, poking and probing it.

WHAT WILL WE SEE?
Reflections from the Black Hole

You surely need no prompting to predict that we won't see the black hole.

Just to prove a point, we'll turn on the ship's powerful outside spotlights and aim them directly at the center of the area where the black hole should be, but can't be seen. The idea is that the light from the spotlights will reflect off anything that's in the hole and we'll see the reflected light. You can predict the result of this experiment. Quite right. Nothing. Try the laser beam then. Still nothing.

Since the problem could be that there's nothing inside the black hole for light to reflect off, we'll chuck in a small, shiny, chrome garbage bin (full of garbage, we hope; let's not be wasteful). It skims off through space, veers off, and, after a short burst of X rays and other radiation, vanishes. Still nothing.

We know that the particles of light (the photons) from the spotlights and lasers might be reflecting off the bin (or whatever's left of it after tidal effects) inside the black hole. But the photons in that reflection can't possibly move back in our direction at all. They're pulled to the singularity in the center of the black hole.

The Black Hole as a Lens
On the viewing screen where we watched the garbage bin disappear, we see what appears to be an ordinary starry sky. Why not an empty black spherical area where the black hole is?

The answer is that a black hole acts like a lens. You learned in Chapter 2 how any massive body bends the paths of light coming near it, and how we can be fooled by this bending into thinking a distant star's position is different from what it really is.

A black hole can bend the paths of light from stars so that we see a normal, starry sky. But it isn't at all normal, as we'll see if we study it carefully and compare it with previous photographs of the same area of sky. There are distortions. We see many stars twice and in the wrong places. Some stars appear farther away than they really are. This happens when the black hole bends the paths of their light so much that the light swings around the hole many times before coming on to meet our eyes. We even see some stars that are behind us!

For a brief instant a circle of light flashes on the viewing screen. An **Einstein ring.** Albert Einstein predicted that if a star or galaxy is exactly centered behind a black hole, the paths of some of its light will be bent around all sides of the black hole at once. The starlight will reach us as a ring of light.

WHAT CAN WE MEASURE?
Black Holes Have No Hair

Can we measure something we can't see?

Having met its fate, the former star has been reduced to no other values than its mass, angular momentum, and electric charge. These three, and nothing else, form all the interesting and complicated features of the star and anything that has fallen past the event horizon.

How can we measure these values?

Realistically we have to expect some mass was carried off in the turbulence of the collapse. However, whatever matter was left when the event horizon formed is still there in the black hole, compressed to disembodied mass at the singularity. Any planet that was orbiting around the star before the collapse might still be orbiting in the same way after the collapse. Think back to the description of the earth being squeezed (pp. 28–29) and you'll understand why this is true.

If we know the angular momentum of the star, we are probably close to knowing the angular momentum of the black hole, because, as we've learned, angular momentum is "conserved"; it can't disappear. How-

Computer simulation of a galaxy
as it would normally appear

*The same galaxy as it might
appear if a black hole were to come
between it and us. The paths of light
coming from some of the stars
in the galaxy are bent around
all sides of the black hole at once.
They reach our eyes as a bright ring,
known as an "Einstein ring."*

ever, it can be carried off or transferred to other objects. Some of this would have happened during the collapse.

When it comes to charge, the star, made of atoms, probably had equal amounts of negative and positive charge. As it collapsed, these would have canceled one another out and left the black hole with a charge of near zero.

Are we sure we can't measure some other things, like the chemical makeup of the star, for instance? No, we can't. No black hole is going to divulge any other secrets about the star it once was. The curtain has come down irrevocably on that star, and every other trace or clue is lost forever. John Wheeler, who made up the name "black hole" in 1969, likes to say, "Black holes have no hair." He's drawn an assortment of items going into a black hole. Leaving what? Mass, angular momentum, charge (Figure 24).

But suppose we hadn't known about the star until after it collapsed to a black hole. Could we measure the black hole then? This becomes a vital question as we search for real black holes in the universe.

One way of measuring the mass of a black hole is to see what it does to objects near it.

We can calculate the **solar mass** if we know the distance and velocity of the planets. We can figure the mass of a black hole by watching orbiting objects— planets or our spacecraft, for instance. We'll see in Chapter Five how useful this is in studying a **binary system,** when a suspected black hole and a visible star are orbiting around one another.

Knowing the mass, we calculate the size of the black hole. You can do this yourself by figuring a radius of roughly 2 miles (3 kilometers) for each solar mass. A 100-solar-mass black hole has a radius of about 200 miles (300 kilometers). Remember, we're

*John A. Wheeler, one of the leading
gravitational physicists of this
century, lectures to his students
at Princeton University. He coined
the term* black hole.

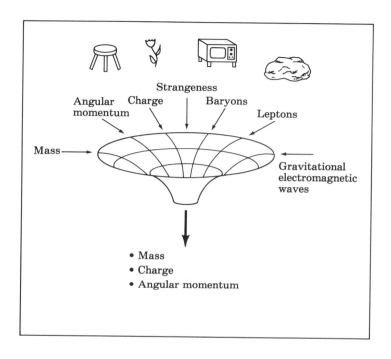

Figure 24. Things falling into a funnel

talking about the size of the black hole, bordered by the event horizon, not the size of the star. It went on shrinking within the event horizon to a dimensionless singularity.

Another way of finding the mass and size would be to study the **lensing effect** of the black hole. How much of the background sky is affected, and in what way?

To measure the angular momentum and rate of rotation of an existing black hole, we can send down a series of test objects (or study matter already circling the black hole) and find out where things begin to be swept around irresistibly with the hole. That will tell us where the ergosphere begins. From the dimensions of the ergosphere we can figure the rate of rotation. You'll recall that only rotating black holes have ergospheres. No rotation, no ergosphere. The greater the rotation, the larger the ergosphere.

Or we can send gyroscopes into the ergosphere. The way they react to the gravity of the black hole will indicate the rate of rotation.

If we were to go around and measure all the black holes in the universe, we'd find that they are almost boring in their sameness. They come in only two basic shapes, spherical or slightly elliptical.

Measuring the Entropy of a Black Hole

Theoretical physicists love to talk about **entropy.** Entropy is the amount of disorder there is in any system—in the universe, for instance, or in a black hole.

There's a rule that entropy can only increase, never decrease. Things never get more orderly. If you clean your room, that area probably does get more orderly, but the expenditure of your energy to do it increases the disorder of the universe by converting that energy into less useful energy, which seems an excellent excuse for never cleaning anything.

When disorder increases, things get less predictable. This loss of predictability means that we can say that in one sense entropy is *unavailable information,* and an *increase* in entropy (disorder) represents a *loss* of information.

A black hole hides information. It reduces it to disembodied mass, rate of rotation, and charge. Anything going into a black hole adds to the amount of lost information. It also adds to the mass of the black hole and the area of the event horizon. We can say that measuring the area of a black hole's event horizon is the same thing as measuring its entropy.

Using this line of thinking, in 1970, Jacob Bekenstein, then a graduate student at Princeton, claimed that black holes have entropy. Nobody wanted to believe it.

Why not?

If something has entropy, it has a temperature. If it has a temperature, it's emitting radiation. If it's emitting radiation, we can no longer say nothing's escaping.

Can We Take a Black Hole's Temperature?

We can't stick a thermometer down into a black hole. When we talk about the temperature of a black hole, we mean the temperature at the event horizon. This has to do with the amount of energy the black hole is radiating.

Got to be pretty cold! you're thinking. Black holes can't generate energy. Nothing can escape past the event horizon!

Common sense says that if nothing can come out, the black hole can't possibly generate energy. The temperature at the event horizon ought indeed to be very cold. However, as we've seen, common sense isn't always the best guide to understanding black holes. We've not yet believed all of our six impossible things before breakfast.

Stephen Hawking of Cambridge University in England is capable of some very uncommon sense. He was disgusted with the idea of a black hole having entropy. He set out to disprove it and instead discovered to his amazement that a black hole, even one that isn't rotating, can indeed generate energy. "A black

78

Stephen Hawking, one of the great theoretical physicists of this century and the discoverer of Hawking radiation, in his study at the Cambridge University, England

hole can shine white hot," as he puts it. It will create and emit particles at a steady rate.

Simplifying Hawking's theory is bound to distort it to a certain degree. Nevertheless, here's a way of picturing it: We think of space as a vacuum, but that's wrong. It's never a complete vacuum. Pairs of particles are continually created there, everywhere—pairs of photons, gravitons—even **pairs of matter particles.** The two particles in a pair start out together, then move apart. Then after only a split second of time they come back together again and annihilate one another. A short but eventful life, particularly at the end.

These are not "real" particles that we can detect with a particle detector, but don't get the idea that they're imaginary. We can measure their effects on other particles. We call them **"virtual" particles.**

You may have learned that the total amount of energy in the universe always stays the same. There can't be any suddenly popping out of nowhere. How do we get around that rule with these newly created pairs? They're created by a very temporary "borrowing" of energy. Nothing permanent at all. One of the pair always has positive energy. The other has negative energy. The two balance out. Nothing is added to the total energy of the universe. In the case of the matter particles, one of the pair will be **antimatter,** a favorite item for science-fiction writers (it drives the starship *Enterprise*), but not purely fiction.

Stephen Hawking points out that there will be lots of these particle-**antiparticle** pairs near the event horizon of a black hole. Let's say that before the pair of virtual particles meet again and annihilate each other, they start to fall into the black hole. The one with negative energy crosses the event horizon. The one with positive energy is still outside.

If, like Pinocchio, a virtual particle dreams of becoming real, the star it wishes on should be a black hole. The gravitational field at the event horizon of a black hole is strong enough to do an astounding thing to virtual particles, even those poor guys with negative energy. It can change them from virtual to real.

The transformation from virtual to real makes a big difference to the pair of particles. The particle with negative energy doesn't have to find its partner and be annihilated. They can live much longer, and separately. The particle with positive energy might fall into the black hole too, of course, but it doesn't have to follow its partner. It's free from the partnership; it can escape. To us, it will look as though it came out of the black hole. It really came from just outside. Meanwhile, its partner will have crossed the event horizon, carrying negative energy into the black hole (Figure 25).

Unfortunately, if we try to measure such radiation from a 100-solar-mass black hole, we'll be disappointed. This is entirely the wrong sort of hole. The surface temperature of a 3-solar-mass black hole would be less than a millionth of a degree above absolute zero. It would be even colder for larger black holes. Stephen Hawking says that our black hole might emit a few hundred photons a second, but they would have a wavelength the size of the black hole, and so little energy we wouldn't be able to know they were there. The rule is, the greater the mass, the greater the area of the event horizon. And the greater the entropy, *the lower the temperature*. You were quite right if you thought it would be cold.

If we were studying a tiny black hole, one the size of an atomic particle for instance, it would be more

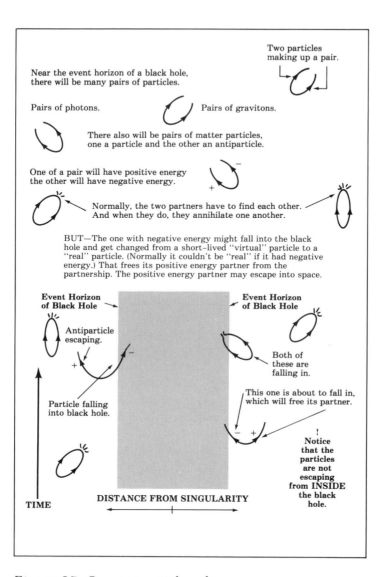

Figure 25. Games particles play.
Hawking radiation

hopeful. Stephen Hawking thinks those little guys absolutely crackle with radiation, which scientists aptly call **Hawking radiation.**

For just a moment, let's consider what such a tiny black hole would be like. Can it have formed from the collapse of a star? Certainly not. To say that it's below the Chandrasekhar limit would be putting it mildly. These **primordial** black holes, as we call them, must be relics from the very early universe. They were never stars at all.

How were they formed then?

We might be able to make one ourselves if we could find a way to press matter together tightly enough. We don't know how to do that, but in the very early universe there were pressures that could. Sometimes it was only a small amount of matter that was compressed. In any case, a primordial black hole will by now be much smaller even than it started out. It will have been losing mass for a long time.

Losing mass? Yes. According to Hawking, a black hole *can* lose mass. As the black hole transforms the virtual particles to real particles, it loses energy. How can this happen, if nothing is escaping through the event horizon? How can it lose anything? It's really rather a trick answer. When the particle carries *negative* energy into the black hole, that makes *less* energy in the black hole. Negative means "minus," which means less.

You can see that Hawking's radiation robs the black hole of energy. But what about mass?

The fact is that when something has less energy, it automatically has less mass.

Remember Albert Einstein's equation $E = mc^2$? The E stands for energy, the m for mass, the c for the speed of light. When the energy (on one side of the

83

equals sign) grows less (as it's doing in the black hole), something on the other side of the equals sign grows less, too. It can't be the speed of light, because that can't change. It has to be the mass (Figure 26). So when we say a black hole is robbed of energy we're also saying it's robbed of mass. This is the only way we know that a black hole can lose mass and the area of its event horizon can grow smaller.

The process will have rather drastic consequences for a primordial black hole. As the mass grows less and the black hole gets smaller, the temperature and rate of emission of particles at the event horizon increase. The hole loses mass more and more quickly. The lower the mass, the higher the temperature. A vicious circle!

Nobody's sure how it ends. The best guess is that the little black hole will disappear completely in a huge final puff of particle emission, like millions of hydrogen bombs exploding. Might that eventually happen to a 100-solar-mass black hole, the size of our imaginary one? The universe will have come to an end long before our black hole reaches that stage.

WHAT CAN WE LEARN?
Blitzing the Black Hole

We know something already about the way a black hole affects things around it. Some of these effects created hazards for the spacecraft. Others helped us measure the black hole. What other experiments should we run?

We'll continue chucking things into the hole. Already we would have noticed a lovely burst of gravity waves and other radiation as we threw in the garbage bin to test for a reflection. We want to measure similar radiation from a number of test objects—objects of different masses and perhaps different electric

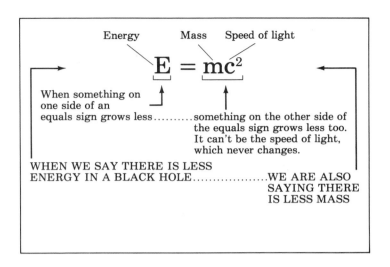

Figure 26.

charges, including microscopic test particles, sent down at many different angles. These experiments may also help us understand the ergosphere, tidal effects, time dilation, and redshift.

For instance, let's watch a small object fall toward the hole. According to theory (no gravity wave or gravitation has yet been directly detected), somewhere still outside the event horizon the object should begin to emit gravity waves more and more strongly. As it approaches the event horizon, the redshift drastically cuts down on the frequency and energy of the gravity waves. We won't observe any emission from the object after it crosses the event horizon. A gravity wave can't escape from within the event horizon any more than a light wave can (Figure 27).

It will be particularly interesting to track test items

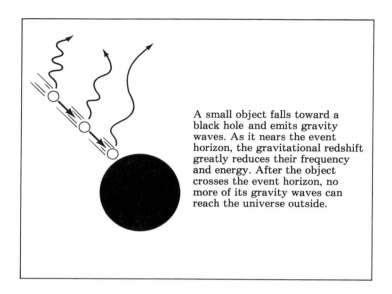

A small object falls toward a black hole and emits gravity waves. As it nears the event horizon, the gravitational redshift greatly reduces their frequency and energy. After the object crosses the event horizon, no more of its gravity waves can reach the universe outside.

Figure 27. Object falling into a black hole

that circle the black hole many times before falling in. This will show us how the black hole forms an **accretion disk.** Accretion disks are important clues in the search for real black holes.

One of the ways we identify black hole candidates in space is by looking for sources of radiation: X rays, **gamma rays,** infrared rays, and radio waves. We know that matter falling toward an event horizon gets stretched and torn apart by tidal effects. It will end up in the form of gas. There are violent collisions and a lot of friction between the molecules in the gas as they orbit the hole. This causes them to lose energy and fall to lower orbits. The lost energy heats everything up, and a lot of radiation is emitted.

Remember, however, that the black hole itself isn't emitting this energy. The radiation comes from the molecules of gas spiraling into it, still outside the event horizon. The spiraling gas forms a disk—an accretion disk—that moves around the black hole like a fat record on a phonograph. Stationed near a black hole, we can study the formation of such a disk and the radiation it emits (Figure 28).

Will the black hole in turn be affected by things around it?

Yes. We'll find the event horizon increasing in area as matter falls in and adds to the mass. We might find a way of studying how the gravitational pull of other bodies raises tides on the event horizon, and how this slows the hole's rotation. On the other hand, enough matter spiraling into the black hole could speed up the rate of rotation, like spinning a top.

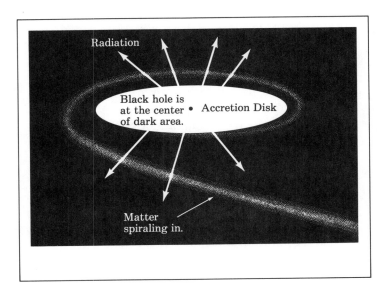

Figure 28. Accretion disk

Some of the sporting members of our crew might like to test their skills in cosmic marksmanship. In something called the **Penrose process,** discovered by Roger Penrose of England in 1969, two particles can perform an exchange of energy within the ergosphere. This results in one of them escaping into space with more energy than it had originally, while the other orbits into the black hole, carrying negative energy with it—a little like Hawking radiation, but not the same thing. You have to have a rotating black hole for the Penrose process to work.

The first object of the game will be to slow down the hole's rotation by placing particles with exactly the right speed and direction in the ergosphere. Each escaping particle carries away some of the angular momentum and mass/energy of the black hole (as negative energy goes into the hole with the doomed particle). Ultimately the hole might have no angular momentum left at all, and stop spinning, but we'll be unlikely to take things that far. In any case, the hole has a certain irreducible mass below which it can't go.

Next we'll try to "spin up" the hole again, by shooting particles into it off-center, giving it back the angular momentum that was carried away. These entering particles will also of course restore mass/energy to the hole.

We'll find it's very tricky extracting mass/energy and angular momentum from a black hole, and even trickier getting a black hole back exactly to its original mass/energy and angular momentum. Normally, if you restore all the angular momentum, you end up with more than the original mass/energy. Only with precisely correct choices of particles (how massive they are), direction of motion, and points of impact can we manage to reverse exactly the changes we've made. It will be a true test of skill in calculation and marksmanship.

One last experiment. Let's place extremely accurate clocks in orbit at various distances from the hole. By comparing them, we'll test theories about time dilation.

Suppose we prove that time comes to a stop (from our point of view) at the event horizon. Should we perhaps build a place to live just outside it? If we did, could we prolong our lives? Could we become immortal?

In one way of looking at it, by our present standards, yes. Not in the best of health perhaps! Don't forget tidal effects. Would it be worth the trouble and discomfort? Surely by now you know the answer.

Things wouldn't seem any different to us. Our lives wouldn't seem any longer. Minutes, hours, days, years, and decades would pass just as they do now. The only way we could know how slow our time was moving would be by seeing ourselves from somewhere else in the universe.

But perhaps . . . capture a tiny primordial black hole . . . keep it in the dresser drawer . . . let everybody wonder how I can stay seventeen while they get old. . . . Oh well, that's probably only the fourth or fifth impossible thing before breakfast!

EVIDENCE IN THE
CASE FOR BLACK HOLES

*If it isn't a black hole, it really has to be
something exotic!*

—*Stephen Hawking*

Before the mid-1960s, many scientists and mathematicians thought that neutron stars and black holes were best left to science fiction. They were amusing to speculate about. They made fiendish mathematical exercises to assign to graduate students. And it had to be admitted that some solutions to Einstein's equations made black holes seem inevitable. But, really, all of it ought to be taken with a healthy dose of skepticism.

How is it then that twenty-five years later we do take them quite seriously indeed? What concrete evidence is there that such things exist?

In 1967, Jocelyn Bell, a graduate research student at the University of Cambridge in England, was working on a project using receivers set up in rows in a field. Through these "beanpoles" she detected mysterious regular pulses of radio waves coming from certain points in space.

The discovery was a chilling one. Could the pulses be coded messages from an alien civilization? Jocelyn Bell and her adviser, Tony Hewish, gave that possibility serious thought but then had to discard it. Why? An alien civilization would surely live on a planet, not a star. A planet moves in orbit. The source of these pulses was not moving in a pattern like an orbit.

By far the most likely explanation was a star that was spinning at a furious rate, many times a second, and sending out a narrow beam of radiation across space like a beacon from a lighthouse. The beam sweeps the earth with each rotation. The original name for these spinning, pulsing stars—LGM (Little Green Men)—was changed. Too bad for us science-fiction enthusiasts! The name pulsars stuck.

Pulsars turned out to be tiny, extremely dense stars only about 20 miles (30 kilometers) across. Jocelyn Bell had discovered a neutron star. This was great news to those few who believed in black holes in 1967. If stars could collapse to such a size and density, was it so absurd to think that other stars could collapse even further and become black holes?

Since 1967, we've discovered more and more neutron stars. Now we're increasingly certain we've also discovered some black holes.

Sometimes, but not very often, in the history of science, we've developed a complicated theory as a mathematical model—a model that we almost have to believe because it makes sense—without any solid evidence from observations to back it up. In other words, an idea works out mathematically and makes beautiful logic, and the more we toy with it the more logical it seems—and nobody can prove it *isn't* true—but we can't find any proof in the real world that it *is* true.

Such was the case with black holes. The theories themselves, in all their detail and elegance, were highly convincing, but until very recently, we couldn't find anything with our telescopes about which everyone could nod in agreement and say, "Now, that *has to be* a black hole."

Even today, not everyone agrees that the theories are highly convincing. Here are some questions skeptics ask:

Are we certain that nothing can stop the collapse of a star after gravity overcomes the exclusion principle? Someone may discover another opponent that competes with gravity.

What about within the event horizon? No one claims to know what happens there. Does the collapse stop there for some reason that hasn't yet occurred to us, before the star reaches the point of singularity? Perhaps gravity self-destructs: if it swallows light, will it not also swallow itself? The future study of how gravity operates on the level of the very small may answer these questions. Maybe we'll find out that collapsing matter never *does* reach a singularity. However, that would not rule out black holes. If an event horizon forms, we have a black hole, even if there is no singularity.

Can we completely trust the theory of relativity? Predictions about the formation of black holes depend on this theory. It tells us that mass curves spacetime and the larger the mass, the greater the curving. We know that the theory has been tested in less drastic situations and it stands up to testing. But we also know that in the ultimate drastic situation, where spacetime curvature becomes infinite, at a singularity, the theory itself breaks down.

What are you and I to think? Are there black holes in spacetime? To be quite frank, doesn't it seem like

92

nonsense that something as large as many many times the mass of the sun will collapse to an infinitely small point? Can we, even in our wildest imaginations, visualize such a thing? Both the largeness and the smallness of it are inconceivable.

In spite of all this uncertainty, the fact is that if some researcher were suddenly to prove that black holes don't exist, it would cause an upheaval of awesome proportions in the world of physics and send some of our greatest living physicists and mathematicians back to square one!

What is the observational evidence for the existence of black holes?

There are things that we see happening in the universe that will be very hard to explain if there are no black holes. There are questions that are very difficult to answer except by saying "There is a black hole there."

For example . . .

WHO ARE THE INVISIBLE COMPANIONS?

Not all stars lead such lonely lives as our sun. Often we find two stars, linked by each other's gravity, orbiting around their common center of mass. We call a partnership like that a **binary system.**

Sometimes, instead of two stars, we detect one, mysteriously moving around exactly as though it were in orbit with a partner. When a star behaves like that, we can be certain it isn't alone out there. It has an invisible companion star. Our own galaxy has many such binary systems in which one of the pair is invisible.

Is the unseen companion a black hole? Not necessarily. It may be a small, dim low-temperature star, perhaps a white dwarf or neutron star. We know that some invisible companions are pulsars. Or it may be a larger star hidden by dust or other debris.

In a few cases, the mysterious companion can't be explained away quite so handily. When satellites orbiting outside our atmosphere detect a lot of X-ray radiation coming from the unseen star, we know there's no dust or debris in the way, because X-ray radiation can't penetrate dust very easily. The star must be small; otherwise we'd see it. It might be a small, dim star—a white dwarf or a neutron star—or it might be a black hole. Knowing its mass would help us decide which.

Is it possible to find out the mass of the unseen star? Yes. Light reaching us from a binary system varies a little. When a star's orbit is bringing it toward us, the Doppler effect isn't the same as when its orbit is taking it away from us. The difference is small, but it's detectable. Furthermore, the brightness of some binary systems increases and decreases, and we conclude that one of the two stars is regularly eclipsing the other. By studying these changes and finding the size and orbits of the system, we find the mass of the system and of each individual star, even when one of them is invisible.

If the mass of the unseen companion is several times the mass of the sun (above the Chandrasekhar limit or the limit for neutron stars), we know we haven't got a white dwarf or neutron star.

An invisible companion with large mass and large amounts of X-ray radiation getting through to us: when we find that, we begin to think "black hole."

Let's take a look (if it's fair to call it a "look") at the candidate in the universe that experts are most certain is a black hole. It's in a binary system in the area of the sky known since ancient times as Cygnus, the Swan. We call it **Cygnus X-1.**

94

The English physicist Stephen Hawking and the American physicist Kip Thorne have a bet on about whether Cygnus X-1 will turn out to be a black hole. Hawking has now decided it is definitely a black hole, and wants to pay off the bet. Thorne says he isn't certain yet, and he won't accept Hawking's payment. How would you place your own bet? Here's the available evidence:

The binary system is in our galaxy, about 7,000 light-years from earth. The visible star is a blue giant.

Studies of the Doppler shift show us that this blue giant has an invisible companion star and that the two stars complete one orbit in 5.6 days. The system is 20 million kilometers across. Knowing this orbital period and size, we figure that the invisible object has a mass between five and ten times the mass of the sun.

Hot gas is flowing from the visible star toward the unseen companion. The X-ray radiation tells us that this gas heats to perhaps a billion degrees Celsius as it approaches Cygnus X-1, forming an accretion disk. Such radiation couldn't penetrate a veil of dust capable of hiding a large star from our sight.

A star much too small for us to see, but having a mass of 5 to 10 solar masses. What else could it be but a black hole? How can there be any doubt?

Remember what you've read about the masses of collapsing stars: A star less than about three times the mass of the sun probably won't become a black hole. It will become a white dwarf or a neutron star. The greater the mass above that limit, the more likely the star will collapse to a black hole.

If we found a binary system in which the invisible companion was 100 solar masses or some such large number, no one would quibble. The problem is that Cygnus X-1 and some of the other invisible companions have masses close to the dividing line between

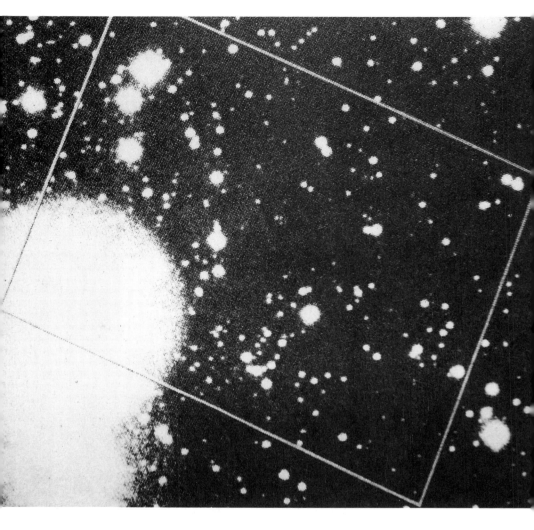

A blue giant star, the visible companion of Cygnus X-1. The rectangle frames the region where the black hole (Cygnus X-1) is thought to be lurking.

*X rays coming from the mysterious
Cygnus X-1 source, which are not
visible through any optical or radio
telescope on the surface of the earth.
What you see is not an object, but
the X-ray radiation coming from the
area within the rectangle in the photo on
facing page. This image was obtained by
the Einstein Observatory, a spacecraft
outside the earth's atmosphere.*

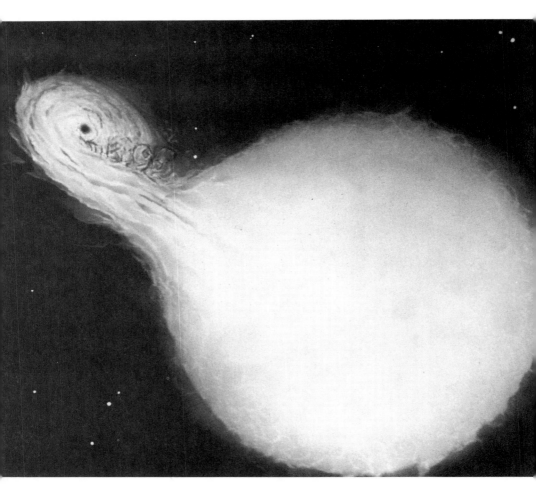

Artist's conception of Cygnus X-1
and its visible companion star.
Matter torn off the large, bright
visible star (right) is drawn
into an accretion disk around the
black hole. There, the gas becomes
superheated and emits X rays before
falling into the black hole.

neutron stars and black holes. This dividing line is not clear-cut. Centrifugal force and magnetism must be considered. No one is yet sure how much or how little these affect the collapse of a borderline star, but they must surely have some influence. Skeptics feel we might find that such stars as Cygnus X-1 are only neutron stars after all, on the brink of collapse but saved by their centrifugal force or magnetism.

WHAT IS THE DARK MATTER?

When we study the universe we come to the conclusion that for everything to work as it does, there's got to be much more matter around than we detect with our present technology. There's evidence that between 90 and 99 percent of the matter in the universe is not radiating at any wavelength. It's completely invisible with any equipment we have. We suspect the presence of this so-called **dark matter** because of its gravitational effect on matter we do see. Of course, we might have made a colossal error in figuring how gravity operates, but that suggestion isn't popular.

Take the way stars move in our Milky Way galaxy. All of them, including our own sun, orbit around the core of the galaxy. In any orbiting system, like our solar system, we expect to find that objects farther out from the center orbit much more slowly than those nearer the center. This ought to be true of the stars in the galaxy. It isn't.

What we know about gravity tells us this can't be right.

The only possible explanation is that there's more to our galaxy than meets the telescope. Much more than the giant, rotating pinwheel with a massive, bulging core and less massive arms.

What would it take to make the difference, to make the stars orbit as they in fact do?

If the matter of the galaxy were mostly outside the core and the disk of the galaxy, if it were to extend well beyond what we see as the edge of the disk, if much of it weren't level with the disk at all but out to the sides of it, then the orbits would make sense (Figure 29).

The spiral shape of the galaxy adds to the puzzle. Computer simulations of galaxy formation can't form a spiral-disk galaxy without figuring in a halo of invisible matter out beyond the matter we can see. It won't work any other way.

By means of radio astronomy we discover what appear to be cold hydrogen clouds extending beyond what we see as the edge of many galaxies. Does this discovery reveal the dark matter? Far from it. It only adds to the mystery. The gas at the outer edges moves far too rapidly for the gravity of the galactic core to hold it in the galaxy. Nothing we observe with optical or radio telescopes can explain why the gas isn't thrown off into space. Invisible matter, making the galaxy much more massive than it seems, is the only possibility.

The evidence begins to pile up:

Polar ring galaxies are galaxies that look like a sphere of stars surrounded by a large ring of other stars. Nothing in our physics can explain how the ring stays in place unless there's more invisible matter out beyond it to balance it.

In 1986, the Canadian Sidney Van den Bergh found that the galaxy known as M87, a giant elliptical galaxy, is surrounded by a cocoon of extremely dim filaments of light, probably stars and hot gas. Had he discovered "dark" matter? On the contrary, there is much too little light to account for the mass needed to keep such gas and stars from escaping.

Some of the most recent evidence has to do with the bending of light from distant sources. As we've

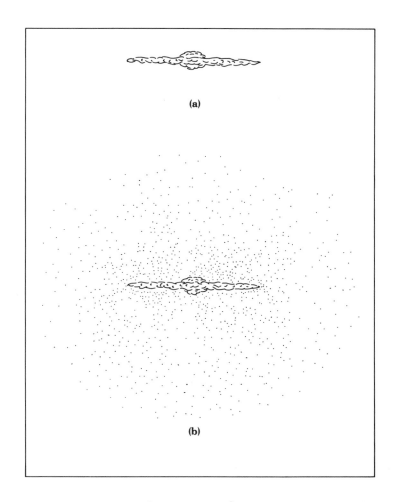

Figure 29. Our galaxy as we observe it (a) and as it must in fact be (b)

seen, the paths of light through spacetime are bent by massive objects. The **bender** (the object causing the bending) may be an entire galaxy rather than a single body. When the bending is far too severe to be caused

101

by the visible matter in the bender galaxy, we again must suspect that we aren't seeing everything that's in that galaxy.

There is larger structure to the universe than galaxies. There are clusters of galaxies. How is it that galaxies in clusters have enough mass to hold together in the cluster without at the same time radiating much more light than they in fact do?

Perhaps the strongest and most interesting argument for the existence of dark matter comes from the study of the early universe. The model many physicists favor today for explaining what happened during the first split second is the "inflationary" model of the universe. If this model is correct, the amount of matter in the universe today ought to be many, many times what we observe.

All right, let's say there *is* all this dark, mysterious matter out there. Most scientists now agree that it exists. Clearly, if we could discover the missing matter, it would solve a lot of our problems in understanding the universe. But what is it and where is it hiding?

Could it be very faint, ordinary stars? Not likely. Telescopes are now sensitive enough to detect such stars in many galaxies. In any case, we'd certainly see more of them lurking about in our own galaxy if they were the dark matter. We're not talking about a wisp of matter here and there. Quite the contrary. If we're to account for the way our galaxy rotates, we have to figure that there's much more invisible matter in it than there is visible matter. Some of the dark matter has to be near enough to the earth for us to detect if it's giving off any radiation to speak of.

Could the dark matter be the hot gas we observe with radio astronomy? There isn't enough of that around.

Could it be planets that are emitting no visible

light? We'd probably detect their infrared radiation if they were beyond the outer edges of a galaxy near us—which is where much invisible matter should be.

We can't escape the suspicion that dark matter may be matter hidden in black holes. In Chapter Four you read about primordial black holes of microscopic size formed in the intense heat and pressure of the very early universe. These black holes would have small masses when compared to those formed from stars, much too small for us to detect as parts of binary systems or benders of light by themselves. However, there may be vast numbers of them scattered all over the universe, especially around the edges of galaxies, enough to account for much of the mysterious dark matter. Would we know if there were one near the earth? Maybe not, unless it reached the end of its life and exploded. Whether or not we should hope for this kind of observational proof is a sensitive question!

Again we must be fair and admit that there are other candidates besides black holes to explain the dark matter. There are types of unusual stars that give off very little light. **Brown dwarfs,** for instance, are extremely difficult to detect. Low-mass stars would be difficult to detect with instruments we have at present. Undiscovered dwarf galaxies are another suggestion.

Other candidates for making up the missing matter are elementary particles, fundamental bits of matter on the size scale of electrons and photons. Neutrinos, for instance, are elementary particles that outnumber electrons and protons in the universe by about a billion to one. Their mass is unimaginably tiny; in fact there is some doubt as to whether they have any mass at all, but in the early 1980s many physicists thought they might add up to form the dark matter. Since then, that possibility has been ruled out. Researchers are now looking for particles called "wimps" (weakly interacting massive particles). None have ever been observed, but some people think wimps could be the dark matter.

So far, we've talked mostly about matter forming halos or cocoons around the edges of galaxies. As we'll see next, a great amount of matter may be trapped in supermassive black holes at the galactic cores.

WHAT IS THE MACHINE IN THE MIDDLE?

When you look toward the constellation Sagittarius in the night sky, you're looking toward the heart of our Milky Way galaxy. If it weren't for intervening clouds of interstellar dust, you'd have a spectacular view of billions upon billions of stars, tightly concentrated in and around the great bulge at the center of the galactic spiral. Neither our eyes nor our optical telescopes can penetrate those clouds. However, the center of the galaxy is a very compact source of radio and infrared rays, and with telescopes that study those rays, we have recently begun to probe the innermost few hundred light-years.

What we find there are countless stars, crowded over a million times more densely than we find them in our own part of the galaxy. There are no planets. If there ever were any, they'd surely have been ripped from their orbits and obliterated as their sun-stars passed too close to one another or collided. There are warm clouds rich in dust. There are cooler gases.

At a distance of between 500 and 1,000 light-years from the center, we find evidence that the galactic core is an extremely explosive region. A ring of matter is expanding outward, like a giant smoke ring, the remnant of a titanic explosion that must have occurred at the galaxy's center about 10 million years ago. Such

View in the direction of the constellation Sagittarius. The box indicates the center of our Milky Way galaxy, which we cannot see because of interstellar dust and debris. Even with our largest optical telescopes, we can see no farther than one-tenth of the distance to the center.

explosions may occur regularly. Inside the ring of matter is a pink **plasma** (ionized gas).

Within the innermost 30 light-years there appears to be a colossal cyclone of extremely hot gas swirling around the heart of the galaxy. The eye of this cyclone seems to be a white-hot disk of gas that's spinning even more furiously. The closer to the center, the faster the spin, like water swirling down a drain. At the absolute center, there's an object that is very tiny by galactic standards, an object only a little larger than our solar system, too small to be shown even as a pinprick on the last of the drawings. The mass of this object is, however, about 3½ million solar masses.

In the past ten years we've been able to find out a great deal about what the center of the galaxy is like. However, we're a long way from understanding how it works—how the machinery operates.

For example, what could cause explosions violent enough to send matter equaling a hundred million times the mass of our sun hurtling out toward the far reaches of the galaxy? A supermassive black hole would explain it.

The black hole itself wouldn't emit matter or energy. However, stars coming too near would be torn

Zooming in closer and closer to the galactic center. This artist's conception begins with a "bird's-eye view" from about 300,000 light-years above the disk of our galaxy and ends with a close-up of the vast whirlpool surrounding the innermost few light-years of the galaxy's heart.

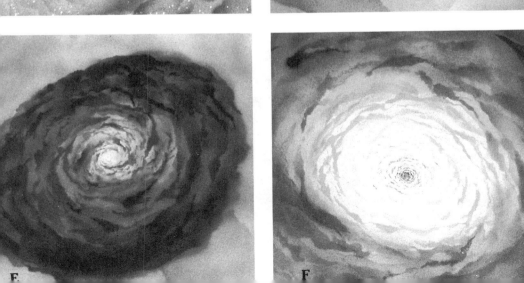

apart by tidal effects. Their remains, and gas pulled off other stars, would spiral inward toward the hole, forming an accretion disk like that around Cygnus X-1, only millions of times larger. Friction would cause heat to build up as gravity pulled the infalling matter into a tiny area, compacting it to unbelievable density. Such a situation could hardly be very stable and smooth. It wouldn't be surprising to find that there were cataclysmic explosions and expulsions of matter every 10 million years or so.

Such a black hole would also explain why the stars at the heart of spiral galaxies revolve as fast as they do. It would also explain why the giant accretion disk isn't broken apart by centrifugal force. In order to have enough gravity to keep the cyclone of gas from flying apart and to cause it to rotate faster and faster as it approaches its center, we need a tremendously massive and compact object, an object of several million solar masses, packed into a region a little larger than our solar system.

There seems to be something at the galactic center about that size. If it's as massive as we think, 3.5 million solar masses, and that small, it must surely be a black hole.

Our knowledge of the galactic core and how the machinery there operates isn't complete. We're still exploring and deciphering clues from the radiation that reaches us. However, experts now believe that black holes are the machines in the middle of galaxies. The case for it is very strong.

WHAT ARE THE LIGHTS AT THE EDGE OF SPACETIME?

Whenever we look out into space, we see the past. Light and other radiation don't reach us instantaneously across space. They travel at about 186,000 miles (300,000 kilometers) per second, and that, of course, is very fast. But even at the speed of light it takes billions of years for pictures of the most distant objects to flash across the universe to our eyes and telescopes and photographic plates. There's no way these pictures can reach us sooner. Like settlers in the old West waiting for the stagecoach, we have to content ourselves with old news.

However, what at first might seem a disadvantage turns out to be a very great advantage. We can look back in time.

The farther distant we look, the farther back we look. The most remote region of the universe in both space and time that we can observe with our present technology now reaches us as a weak hiss of radiation in the radio spectrum. This radiation is a glow from the primeval, white hot, superdense matter of the early universe, 10 to 20 billion years ago.

The next most distant things in time and space that we observe are **quasars.** There are none near us. Quasars are all incredibly far away and long ago, shining to us from the dawn of spacetime. Astronomers at Princeton and California Institute of Technology discovered one in the autumn of 1989, Quasar P.C. 1158 + 4635, whose redshift is so enormous that we see it as it existed when the universe was about a billion years old, only 7 percent its present age.

Are quasars still shining today? If they are, they may by now have evolved into something quite different from the way we observe them, and they are by now much more distant.

What do we observe? Something that looks from earth like a faint star but also in some ways resembles a gaseous **nebula.** Something inconceivably bright that varies in its brightness. Something with a tremendous redshift. Something very small by cosmic standards.

When astronomers Alan Sandage, Thomas Matthews, and Maarten Schmidt of the California Institute of Technology first studied such objects in the early 1960s, it seemed that what they found couldn't possibly exist. If the redshift of the objects was caused by a gravitational field, they would have to be so massive and so near to us that they would mess up the orbits of solar system planets. That wasn't happening.

The other explanation for the redshift seemed equally preposterous. We know that as the universe expands, galaxies and clusters of galaxies move away from our galaxy and from one another. The farther away they are, the faster they're moving. When an object moves away, the Doppler effect causes a redshift. To cause such a redshift as astronomers were finding, some of the mysterious objects had to be flying away from us at a rate near the speed of light. That meant they were very far away indeed.

Could an object be visible at all at such a distance? Was it terribly large? How could we find out?

Quasars vary in brightness. They flicker. Logically, no source of light can flicker any faster than radiation can cross it. Otherwise the next flicker would begin before the last one ended and the flicker would appear blurred to us. We wouldn't see it as a flicker at all. Radiation can't cross a flickering object faster than the speed of light. Timing the flicker and knowing how fast light travels, we find out that quasars are tiny by cosmic standards. Only if they are by far the brightest objects in the universe, as bright as dozens or even hundreds of ordinary galaxies put together, would we be able to see them as we do. Such brightness could only be caused by an energy source far beyond any we can easily explain.

This makes for one of the foremost puzzles in astrophysics: How can anything so small radiate such incredible energy, at all wavelengths, from radio to X ray. The object must be violent indeed. We also know that for it to have a nebulalike appearance there must be huge regions of gas being ionized, superheated, and knocked around violently. If this sounds like the center of our own galaxy, it is. However, what goes on in a quasar is many times more violent. At first, scientists thought that quasars were strange beasts that were going to require a whole new physics to explain.

We know of only one possible source of power great enough to account for the enormous amounts of energy a quasar emits: a black hole of a hundred million or even a billion solar masses. Such a black hole would be tearing whole stars apart and consuming them.

Quasars are too far away and too difficult to observe to provide conclusive evidence. However, it's possible to learn more about them by studying their close cousins, the active galaxies.

People who study galaxies divide them into two broad categories: active and normal. Our Milky Way galaxy is a normal galaxy. There are no great changes taking place in the brightness at its core in spite of the dramatic activity there. **Active galaxies,** on the other hand, flicker as we've said quasars do, changing brightness over days, months, or even years. Their cores undergo violent, energetic upheavals. Gas and stars there appear to be whirling rapidly, as they do at the center of the Milky Way, but so much more violently as to make that look peaceful by comparison.

Through optical telescopes, quasars don't look at all like galaxies. However, in radio pictures quasars and active galaxies turn out to be very similar.

At the core of an active galaxy or quasar is a faint,

tiny source of radio waves. From this, narrow jets emerge, extending over vast stretches of space. At the ends of the jets are huge, puffy, turbulent lobes. The whole thing looks like a great, cosmic, double-ended chicken drumstick.

The process works something like a teapot billowing steam into cold air. At first the steam forms a narrow pillar. Farther away from the spout, the steam meets cooler air, slows down, and forms a turbulent lobe before disappearing. It looks as though it were meeting an invisible barrier.

We now believe that something at the heart of a quasar shoots out hot gas at near the speed of light. Whatever does this also makes a strong magnetic field that acts like a pipe, funneling gas out in the form of the two narrow jets. Far away from the center, the magnetic field grows less and less strong and disappears, and the jets of gas slow down as they meet somewhat cooler gas, the "intergalactic medium," which is much thinner than the most perfect vacuum we can possibly create in a laboratory. The hot gas bunches up and forms the lobes (Figure 30).

We can tell that the engine at the heart of these galaxies and quasars that manages all this isn't very big. The energy and funneling starts off in a space not much larger than our solar system. We know that it's massive, sometimes several billion solar masses. We can see that it's capable of channeling hot gas and delivering it over a time span of millions of years, at nearly the speed of light, to distances over a billion times farther away than the size of the engine's heart.

In 1987, the German astronomer Norbert Bartel and his associates came closer to finding out what the core of an active galaxy is like than anyone had done before. They studied an active galaxy called 3C84 at

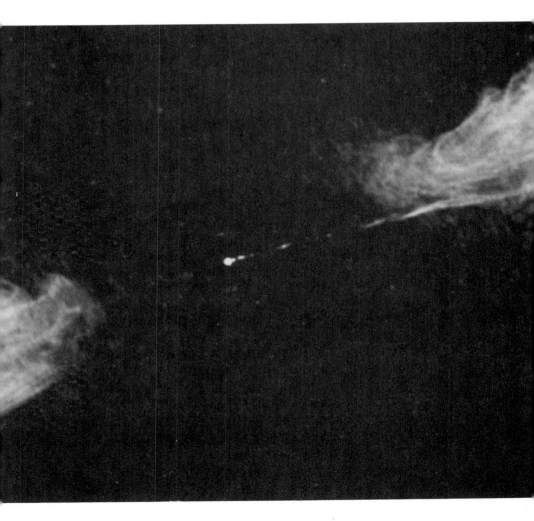

Cygnus A, a "double radio source."
From its core, a tiny point at the
middle of this image, narrow jets
of hot gas extend over vast
stretches of space, ending in
the puffy, turbulent lobes.

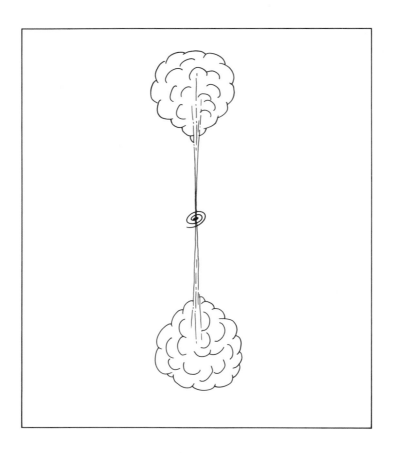

Figure 30. How a double radio source works

such high resolution that they were able to observe a region at its center less than a tenth the size of any studied before. What they found looks somewhat like a sphere surrounded by an extended pancake-shaped ring. It can't be a collection of many stars because there is too much radio emission to be coming from such a small region. More likely it is one compact star more than a million times more massive than our sun. Such a star would almost certainly be a black hole.

As matter spiraled into such a black hole, it would make the hole rotate in the same direction as the spiral and could develop a magnetic field something like that of the earth but far stronger. Before the matter fell in, it would form an accretion disk (the pancakelike ring found by Norbert Bartel) which would emit radiation. Before the in-falling matter crossed the event horizon it would generate very high-energy particles. The magnetic field would be so strong that it could focus these particles into jets ejected outward in the directions of the black hole's north and south poles.

It's beginning to look as though every galaxy in the universe (including quasars, the most active galaxies of all) may have a black hole at its center.

6
WILD IDEAS

I'm entering the black hole now! . . . Good
grief! . . . It's full of unmatched socks!
—from a cartoon by "Chase"

When we're discussing black holes, the fine line between respectable theory and preposterous fantasy
sometimes gets fuzzy.

Everything you've read so far in this book is sup-
ported by current theory. Beyond this there are some
wild rumors going around. If your friends think you
know something about black holes, they're much
more likely to ask "Is there a black hole in the Bermuda Triangle?" or "Is the Loch Ness monster going
in and out of a wormhole?" than to ask at what radius
the event horizon forms.

Here are some far-out, maybe just barely plausible
ideas about black holes—to help you when those dis-
cussions arise.

CLOSE ENCOUNTER AT TUNGUSKA

What if a black hole came near the earth or collided with it? Some people think it's already happened.

For a remote area on the Central Siberian Plateau, not far from the Stony Tunguska River, it wasn't just a close encounter, it was a direct encounter. But with what? A black hole? This idea crops up again and again when black holes are the subject of conversation. Here are the facts:

On the morning of June 30, 1908, a fireball passed across the sky of China and Russia. When it was still several miles above the ground, there was an explosion that disturbed the magnetic field of the earth. Although apparently no one was killed (only reindeer herders go to the area), there were reports of a heat blast bending fir trees and searing clothes and skin many miles away. Witnesses a thousand kilometers southeast saw a cylinder of fire rising far into the sky, and buildings trembled. For many nights, the sky in northern Europe was strangely brilliant. Londoners phoned the police to inquire whether the northern part of the city was afire.

So remote is the Tunguska area of Siberia that it wasn't until nineteen years later that an expedition, slogging through icy swamps and dense forest, made its way to the site.

What did they find? Incredible devastation; trees within a 18- to 25-mile (30- to 40-kilometer) radius ripped up and laid out like charred pickup sticks, their dead roots pointing to the center of the blast area; trees farther away blackened and stripped bare; nothing left uncharred; the ground scorched and riddled with unexplainable holes and ridges; no crater; no metallic remnants at the center of the area such as most meteorites leave.

117

Was it a meteorite? That first expedition decided it was. Some people still question their explanation. Meteorites don't usually explode just before hitting the ground. And why no crater?

A comet? Surely it would have been seen earlier approaching the earth.

Antimatter? As we've learned in the case of particle pairs, when matter and antimatter meet they annihilate one another. But wouldn't the antimatter have encountered matter and met its doom higher in the atmosphere?

Was it a crippled alien spaceship, whose pilot knew its nuclear-powered engines were about to explode and deliberately aimed it for one of the areas in the world where it would cause least harm?

Or was it a tiny primordial black hole? As it approached the ground it would have created a shock wave in the air great enough to produce temperatures from ten to a hundred thousand degrees. The radiated light, absorbed and reradiated by air along the path, would have been the blinding pillar of fire. The heat and radiation would have caused the burning and charring. The blast would have blown down the trees and shaken the earth, even though the black hole itself didn't explode.

The black hole would have passed straight through the earth. (You'll recall that primordial black holes can be smaller than an atom.) On the opposite side of the earth, it would have emerged in open ocean, causing an atmospheric disturbance and an impressive geyser of water as it headed off into space once more.

It's intriguing and not entirely ridiculous to suggest that the Tunguska disaster might have been caused by a black hole, but unfortunately you're going to have to disappoint your friends on this one. Nobody recorded any geyser or atmospheric disturbance. More

118

conclusive yet, in 1980, debris of a comet or meteor was found at the site of impact.

Sorry!

THE MATTER-EATER

If there had been a black hole at the root of the Tunguska mystery, we're extremely lucky that it went right on and passed through the earth and off into space. It might instead have oscillated back and forth through the earth again and again, eventually settling down at the center. Imagine it there; a tiny black hole the size of an atomic particle. What harm could such a little fellow do?

First, it would suck in the particles near it. That would increase its mass and extend its event horizon, putting it in contact with still more particles. Those would be the next victims. It would grow and grow. Eventually, the entire earth would be a black hole. But that didn't happen. We'd know by now—wouldn't we?

Physicists disagree. Some, when asked how long it would take a primordial black hole to consume the earth, answer, "Not long!" Others, perhaps remembering all the empty space within atoms, say, "Longer than the future of the universe."

HAWKING'S RADIATION AND UNMATCHED SOCKS

Some respectable members of the scientific community suggest that black holes in clothes washers are responsible for the distressing disappearance of socks in the washing. Stephen Hawking's work on particle-pair separation at the event horizon sheds new light on this perplexity. It now becomes clear why nearly always only *one* of a pair vanishes.

119

Socks may become "real" in the wash. Nevertheless, even a "real" unmatched sock has little value except as a toy for the dog. Equally useless, except for scientific interest, is an unmatched sock which appears in a household where one of its size, type, color, and odor has never before been observed. In this latter case, the presence of a wormhole must be suspected.

CREATION IN THE BASEMENT

While you're in the basement, puzzling over socks and plumbing, consider the possibility that certain types of instability in our universe might make it "blister" in such a way as to form another universe. The only evidence of this would be a pinpoint in our universe where the curve of spacetime was infinite. There may be a good many of these pinpoints around—singularities inside black holes. Maybe whenever a black hole forms, a new universe is born on the "other side" of spacetime.

Could we create one ourselves? It would take only a tiny bit of matter. With our technology we don't know how to compress matter to such density, but it's not inconceivable that we might learn someday. It's not inconceivable that somebody else learned earlier. Says Alan Guth, of the Massachusetts Institute of Technology, "For all we know our own universe may have started in someone's basement."

THE ULTIMATE GARBAGE RECYCLER

Could we put a black hole to work for us? Charles Misner, Kip Thorne, and John Wheeler make this proposal (see Figure 31):

An advanced civilization builds a rigid framework around a rotating black hole, not right at the event horizon, not even at the static limit; in fact, pretty far

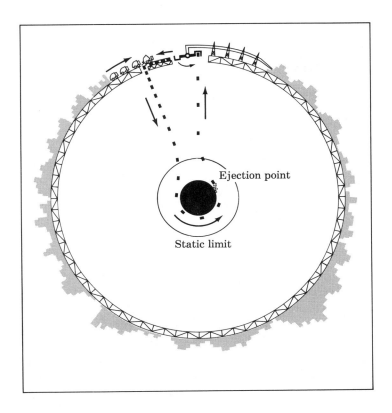

Ejection point

Static limit

*Figure 31. The ultimate
garbage recycler*

out from both. On this framework, they build a city.
Every day, garbage trucks carry all the city's garbage
to a dump. There the garbage is shoveled into shuttle
vehicles, and these are dropped toward the black hole
through an opening in the rigid framework.

A garbage shuttle reaches the static limit and en-
ters the ergosphere. There, it gets whipped into a circling, inward-spiraling orbit near the event horizon.

121

When it reaches a certain "ejection point" the shuttle ejects its load of garbage. The garbage goes down the hole, increasing the mass/energy of the hole. The shuttle recoils from the ejection and goes flying out away from the hole with more energy than it had going down.

As shuttles come hurtling out through the opening in the rigid framework, they strike a flywheel which turns a generator, producing electricity for the city. The inhabitants of the city in this way use the Penrose process to convert the entire rest mass of their garbage and some of the mass of the black hole into electric power.

THE DONKEY AND THE CARROT

Another suggestion for putting black holes to work comes from Stephen Hawking.

One primordial black hole emits enough energy to run six to ten large power stations. Harnessing its power is the problem. Remember that a tiny black hole on the surface of the earth would fall right through and oscillate back and forth and might end up as a "matter-eater" inside the earth. A better place for it would be in orbit around the earth.

How would we get it there? We could attract it by towing a large mass in front of it, like a carrot in front of a donkey.

Not, Hawking admits, a very practical proposal, at least in the immediate future.

GALACTIC GRAND PRIX

The absolute ultimate for racing enthusiasts and daredevils of tomorrow!

The point of the race: to circumnavigate the galaxy and reach the finish line *last* and *youngest*.

Participants start off in their spaceships. The more black holes they can visit, and the nearer they can skim the event horizons, the more they'll benefit from time dilation and the longer they'll extend their lives. But they must calculate skillfully to avoid lethal tidal effects, and take care never to cross an event horizon (which of course can't be seen), because then they couldn't come back to claim their prize. The best pilots will return to earth billions of years after they left it, still young, athletic, ready to receive the accolades of their remote descendants—who we hope will remember there's a race going on.

But what about the Bermuda Triangle and the Loch Ness monster? If you've read this book carefully, you'll be able to field those questions on your own.

7
GRAVITY IS PATIENT

*My Lord, what a morning . . . when the stars
begin to fall!*
—*American black spiritual*

It isn't the prospect of having clever party conversation or developing power stations that entices some of the earth's greatest intellects into messing around with black holes. Nor is it the prospect of outdoing the White Queen in the number of impossible things one can believe before breakfast—though that's closer.

The most compelling motive is the hope of uncovering something about how the universe works and why, some clue leading to one "deep, happy mystery" that explains every bit of it very simply.

A black hole makes an excellent laboratory for solving the puzzle of the universe because a black hole demonstrates so clearly how gravity operates.

It was the study of black holes that led physicists in the 1960s to conclude that the universe began as a singularity, where density and curvature of spacetime were infinite, which exploded (the so-called **Big Bang**) and began to expand. You can see that this would be a sort of collapse-to-a-black-hole-run-backward.

Black holes help us understand the process of universal expansion—how rapidly it could have happened, how smoothly, how it might have formed clumps and irregularities that later got to be stars, planets, and galaxies. Imagine John Wheeler's funnel in reverse: everything that was to be you and I and galaxy clusters and television sets emerging from the original singularity.

Black holes also help us understand the process of universal collapse. The universe may have begun as a singularity and will go on expanding forever. Or it may have begun as a singularity, is at present expanding, and later will reverse itself and contract again to a singularity. It may also be that spacetime has no boundaries and no singularities any more than the surface of a ball like the earth does (but with more dimensions than a ball).

If it is to contract again, there must be a force already at work that will eventually stop the expansion and do the closing. That force is, of course, gravity. As far as we can tell, it's at work everywhere in spacetime. Every bit of matter in the universe, from subatomic particles to galaxy clusters, feels the gravitational pull of every other bit of matter in the universe. Does it all add up to enough gravitational attraction to end the expansion? Will all the matter begin to draw together, as we saw the atoms do to form a star?

Gravity's a familiar acquaintance, probably the first science we learned as tots. It made us skin our knees and caused ice cream to splat on the floor instead of staying nicely upside down on our spoons. We've all had to learn to live with it and make concessions to it.

We've also managed to defy it in airplanes and space-craft. Even though no one has yet been able to detect a gravity wave or a graviton, and we're still working on a satisfactory quantum theory of gravity, we've understood how gravity works in more or less normal circumstances ever since Isaac Newton in the seventeenth century.

Gravity is indeed commonplace and necessary for our existence. It's also the weakest of the basic forces of nature. But it isn't such a tame beast. We learn from black holes that on the astrophysical scale it can add up in such a way as to overwhelm all the other forces. Press enough matter together and you unleash a monster, powerful enough to extinguish all the incredible detail of a star or a system—everything except mass, angular momentum, and charge. In universal collapse, in the end—the **Big Crunch**—if the universe does become a singularity of infinite density and spacetime curvature, even these three values will surely lose all meaning.

The tug-of-war that gravity plays on the universal level is for high stakes. The game isn't going to be over for a long time. We don't have to have nightmares about the outcome. The mass that makes up our bodies will participate in the crunch, when and if it occurs, but you and I and our world will be extremely long ago and far away.

Gravity is patient.

GLOSSARY

Acceleration—A change in the speed or direction of an object.

Accretion disk—A disk like a fat phonograph record made up of hot gas as it spirals into a black hole. The gas heats to millions of degrees Celsius. We are able to detect the radiation from such a disk.

Active galaxy—A galaxy in which the core is undergoing violent, energetic upheavals and is changing brightness over days, months, or years.

Angular momentum—A quantity related to the speed and mass of a rotating object.

Antimatter—Matter made up of antiparticles.

Antiparticle—Every type of matter particle has a corresponding antiparticle. When a particle meets its antiparticle, they annihilate each other, leaving only energy.

Atom—A unit of ordinary matter. The center of the atom is the nucleus, made up of protons and neutrons. Electrons orbit the nucleus.

Axially symmetrical—Symmetrical with respect to an axis. The same "all round" the axis.

Axis—The imaginary pole down the center of a rotating object, around which the object rotates.

Bender—A massive body or galaxy in spacetime which bends the paths of light passing near it.

Big Bang—The state of enormous heat and density in which the universe probably began, and from which the universe has expanded and cooled to its present state; not necessarily a singularity.

Big Crunch—The singularity which may occur at the end of the universe.

Binary system—Two stars bound by their gravitational attraction. The stars move in elliptical orbits around their common center of mass.

Black hole—An area of spacetime from which nothing, not even light, can escape because gravity there is so strong. This area can't be seen by an outside observer.

Brown dwarf—A star with less than a tenth the mass of the sun which isn't hot enough to transform hydrogen into helium.

Chandrasekhar limit—About one and a half times the mass of the sun. Exclusion principle repulsion among the electrons in a star more massive than this can't support the star against its own gravity.

Conservation of angular momentum—The law in science which says that no angular momentum can appear out of nowhere or disappear.

Conservation of energy—The law in science that energy (or its equivalent in mass) can't be either created or destroyed.

Curvature of spacetime—Einstein's general theory of relativity explains the force of gravity as the way the distribution of mass in spacetime causes something that resembles the warping, denting, and dimpling in an elastic surface by balls of different weights and sizes lying on it.

Cygnus X-1—The invisible member of a binary system in the constellation Cygnus. Physicists now generally agree that it is a black hole.

Dark matter—Matter that we don't observe but think must be present in the universe in order to explain the activity of things we do observe—such as the way galaxies rotate.

Doppler effect—A shift in the wavelength of waves (such as light waves or sound waves) received from an object as it moves toward or away from us. We're familiar with it as the change in pitch as a siren passes us.

Einstein ring—If a black hole or other bender is centered perfectly between us and a background star or galaxy, the light from that star or galaxy is bent around all sides of the black hole at once, and comes to us as a ring of light. The size of the ring can be used to calculate the mass of the black hole or other bender.

Einstein's general theory of relativity—The theory in which the gravitational force is explained as a curvature of spacetime.

Electromagnetic force—One of the four basic forces of nature. The electromagnetic force causes the electrons in an atom to orbit the nucleus. The messenger particle of the electromagnetic force is the photon, the particle of electromagnetic radiation, including light.

Electromagnetic radiation—Radiation made up of electromagnetic waves. Radio waves, visible light, gamma rays, X rays are all forms of electromagnetic radiation.

Electromagnetic spectrum—The breaking up of electromagnetic radiation (including visible light) into component wavelengths.

Electron—A particle with negative electric charge that is most commonly found orbiting the nucleus of an atom.

Entropy—The amount of disorder there is in a system. A law of science (known as the second law of thermodynamics) says it can never decrease.

Ergosphere—The area around a rotating black hole in which anything that is present will be forced to move around the black hole in the same direction the hole is rotating.

Escape velocity—The speed necessary to escape the gravity of a massive body such as the earth and get away to elsewhere in space. Escape velocity for the earth is about 7 miles (11 kilometers) per second. Escape velocity for a black hole is slightly greater than the speed of light.

Event horizon—The boundary of a black hole, which forms when the surface of a collapsing star reaches the radius where light can no longer escape. It isn't possible for anything that can't go faster than the speed of light to escape from inside it. Redshift and time dilation become infinite at the event horizon.

Exclusion principle—The principle which states that two identical particles of matter can't have both the same position and the same velocity.

Free-fall orbit—An orbit not assisted by rocket engines, which depends on nothing except a continual "falling around the earth" (or around whatever it is we're orbiting). Aboard a spacecraft in such an orbit we feel weightless, regardless of how strong the gravity of the body we're orbiting.

129

Gamma ray—Electromagnetic waves of very short wavelength and very high energy.

Gravitational radiation—Energy emitted in the form of gravity waves by such things as collapsing stars.

Gravitational radius—When the surface of a collapsing star reaches this radius, gravity there becomes so strong that nothing can escape, even traveling at the speed of light. You may figure it by multiplying the number of solar masses by 2 for miles or 3 for kilometers. Thus, for a 10-solar-mass star, the gravitational radius will be about 20 miles or 30 kilometers.

Gravitational redshift—The shift in wavelengths toward the red end of the spectrum, caused by a gravitational field.

Graviton—The messenger particle of the force of gravity. It is to the force of gravity what the photon is to the electromagnetic force. Gravitons are believed to travel at the speed of light.

Gravity—One of the four basic forces of nature. Gravity always attracts, and it works on all levels, from the tiniest fundamental particles to the largest objects in the universe.

Gravity wave—When we think of gravitational radiation as waves rather than as particles, the waves are called gravity waves.

Hawking radiation—Radiation produced by a black hole. It can be understood as the appearance of pairs of virtual particles near the event horizon, where one of the partners falls into the black hole and can't rejoin its partner, freeing the partner to escape into space.

Helium atom—The most familiar form of helium atom has two protons and two neutrons in the nucleus, and two electrons.

Hydrogen atom—An atom with one proton in the nucleus (no neutron) and one electron.

Kerr black hole—A rotating black hole. It won't be spherical; it will bulge out around the equator.

Lensing effect—The name for what happens when we see an Einstein ring or other distortion of the sky (such as double images) caused by the presence of a black hole or other bender.

Mass—How much matter there is in a body, or how much a body resists any change in its speed or direction.

Matter—Everything in the universe (and perhaps elsewhere) that is made of matter particles is called matter. Matter particles are the particles found in atoms: electrons, protons, and neutrons, and, on a smaller scale yet, the quarks that make up protons and neutrons. In a more technical definition of "matter," *all* particles are matter, including photons and gravitons.

Nebula—A cloud of gas and dust in space.

Neutron—A particle that is similar to a proton, but with zero electric charge, found in the nucleus of atoms. It's made up of three smaller particles called quarks.

Neutron star—A cold star, about 20 miles (32 kilometers) in diameter. One of the ways a star can spend its old age. Exclusion principle repulsion among its neutrons balances the pull of gravity.

Nuclear fusion—The process by which atomic nuclei coalesce to form heavier atomic nuclei. The fusion of hydrogen to helium is an example.

Nucleus—The central part of an atom, consisting of protons and neutrons.

Particle—Basic building block of nature such as the electron or proton. Some can be broken down into even more basic building blocks: a proton is made up of smaller particles called quarks. We also refer to particles of light or gravity. These are "messenger" particles that carry these forces. Similar "messenger" particles carry the weak and strong nuclear forces.

Particle pairs—Pairs of particles that are being created everywhere in the "empty" space and all the time. They are virtual particles, extremely short-lived, and can't be detected except indirectly by observing their effect on other particles. In a fraction of a second the two particles in a pair must find each other again and annihilate each other.

Penrose process—A process that might theoretically extract energy and angular momentum from, or restore them to, a black hole.

Photon—A "messenger particle" which carries the electromagnetic force. The photon is the particle of light and all other forms of electromagnetic radiation (gamma rays, X rays, radio waves, etc.).

Plasma—Ionized gas.

Polar ring galaxy—A rare galaxy that looks like a sphere of stars within a large ring of other stars.

Primordial black hole—A black hole created in the very early universe.

131

Proton—One of the particles in the atomic nucleus. Protons have positive electric charge and are made up of three smaller particles called quarks.

Pulsar—A neutron star that rotates very rapidly and sends out regular pulses of radio waves, sometimes several hundred to a thousand times a second.

Quantum state—When we say that two particles occupy the same quantum state, we mean that they have identical values of momentum, charge, and spin in the same region of space.

Quasar—An extreme form of active galaxy. Quasars are the most luminous objects known in the universe.

Radius—Shortest distance from the center of a circle or sphere to the circumference or surface.

Redshift—A shift in the wavelength of light (or other electromagnetic radiation) toward the red end of the spectrum.

Schwarzschild black hole—A nonrotating, spherical black hole which has no electric charge.

Singularity—The dimensionless point at the center of a black hole, where all the mass of the collapsing star has shrunk to infinite density. The curvature of spacetime here is infinite. Singularity may also be used to name any point in spacetime where the curvature of spacetime becomes infinite, such as the Big Bang singularity.

Solar mass—The mass of the sun.

Spacetime—The combination of the dimensions of space and time. The spacetime we experience has three dimensions of space and one dimension of time.

Speed of light—Usually when physicists talk about the speed of light, they round it off to 186,000 miles (300,000 kilometers) per second. If you want it more nearly exact, it's 186,300 miles (or 299,800 kilometers) per second.

Static, or stationary, limit—The outside border of the ergosphere around a rotating black hole. Inside this limit nothing can resist being dragged around the black hole in the direction in which the black hole is rotating.

Supernova—The explosion of a star.

Theory of relativity—See Einstein's general theory of relativity.

Tidal effect—The stretching effect caused by the difference in gravitational pull on two parts of the same body.

Time dilation—An effect caused by gravity or the curvature of spacetime, or by acceleration, in which the rate at which time passes in one location will be measured differently by a clock right there and another clock somewhere else.

Time-line—The usually vertical line in a diagram showing the time dimension.

Virtual particle—A short-lived particle which we can't detect directly. We know it exists because we can measure its effect on other particles.

Visible spectrum—The part of the electromagnetic spectrum in which the wavelengths are right for our eyes to receive. Our eyes interpret the different wavelengths in this part of the spectrum as different colors.

Wavelength—For a wave, the distance from wave crest to wave crest.

White dwarf—A cold star, a few thousand miles in diameter. One of the ways a star can spend its old age. Exclusion principle repulsion among its electrons balances the pull of gravity to keep it from collapsing.

Wormhole—After a body falling into a rotating black hole is torn apart and reduced to elementary particles by tidal effects, it may be possible for those particles to miss the singularity, travel through a small hole or tunnel in spacetime, and end up in another universe or another part (or time) of our own universe.

X rays—Electromagnetic waves with short wavelengths and high energy.

SOURCES USED

BOOKS

(*Asterisks* indicate books accessible to young adult readers)

Ames, William L., and Kip S. Thorne. "The Optical Appearance of a Star That Is Collapsing Through Its Gravitational Radius." In *Astrophysical Journal,* vol. 151., February 1968, pp. 659–70.

* Boslough, John. *Beyond the Black Hole: Stephen Hawking's Universe.* Glasgow: William Collins Sons and Co., Ltd., 1985.

* Chaisson, Eric. *Relatively Speaking: Relativity, Black Holes, and the Fate of the Universe.* New York: W. W. Norton & Company, 1988.

* Cohen, Nathan. *Gravity's Lens: Views of the New Cosmology.* New York: John Wiley and Sons, 1988.

* Cornell, James, ed. *Bubbles, Voids and Bumps in Time: The New Cosmology.* Cambridge: Cambridge University Press, 1989.

* Davies, Paul C. W. *The Forces of Nature.* Cambridge: Cambridge University Press, 1979.

——*The Search for Gravity Waves.* Cambridge: Cambridge University Press, 1980.

——*Superforce.* London: Unwin Paperbacks, 1987.

Dodd, James E. *The Ideas of Particle Physics.* Cambridge: Cambridge University Press, 1984.

Fowler, William A. *Nuclear Astrophysics*. Philadelphia: American Philosophical Society, 1967.

Hawking, Stephen W. "Black Holes and Their Children, Baby Universes." Unpublished.

* ———*A Brief History of Time: From the Big Bang to Black Holes*. New York: Bantam, 1988.

———"The Edge of Spacetime." In Paul Davies, *The New Physics*. Cambridge: Cambridge University Press, 1989.

———"The Quantum Mechanics of Black Holes." In *Scientific American*, vol. 236, no. 1, January 1977, pp. 34–40, etc.

Kaplan, Samuil Aronovich. (Translated by Renata Feldman.) *The Physics of Stars*. Chichester, England: John Wiley & Sons, Ltd., 1982.

Matzner, Richard, Tsvi Piran, and Tony Rothman. "Demythologizing the Black Hole." In Tony Rothman, *Frontiers of Modern Physics*. New York: Dover Publications, Inc., 1985.

Matzner, Richard, Tony Rothman, and Bill Unruh. "Grand Illusions: Further Conversations on the Edge of Spacetime." In Tony Rothman, *Frontiers of Modern Physics*. New York: Dover Publications, Inc., 1985.

Misner, Charles W., Kip S. Thorne, and John Archibald Wheeler. *Gravitation*. San Francisco: W. H. Freeman and Company, 1973.

Shapiro, Stuart L., and Saul A. Teukolsky. *Black Holes, White Dwarfs, and Neutron Stars*. New York: John Wiley and Sons, 1983.

———*Highlights of Modern Astrophysics*. New York: John Wiley and Sons, 1986.

Sullivan, Walter. *Black Holes: The Edge of Space, the End of Time*. Garden City, N.Y.: Anchor Press/Doubleday, 1979.

Thorne, Kip S. "Gravitational Collapse." In *Scientific American*, vol. 217, no. 5, November 1967, pp. 88–102, etc.

Waldrop, M. Mitchell. "The Quantum Wave Function of the Universe." In *Science*, vol. 242, December 2, 1988, pp. 1248–50.

Wheeler, John Archibald. *Journey into Gravity and Spacetime* New York: W. H. Freeman and Company, 1990.

Will, Clifford. "The Renaissance of General Relativity." In Paul Davies, *The New Physics*. Cambridge: Cambridge University Press, 1989.

INTERVIEWS

Stephen Hawking, Nicholas Phillips, Sir Brian Pippard, John A. Wheeler; and telephone conversations with David Arnett, Eric Chaisson, Fay Dowker, Andrew Dunn, Stuart Shapiro, Saul Teukolsky, and William Unruh.

INDEX

Acceleration, 51–52, 60–61, 127
Accretion disk, 86, 87–95, 98, 108, 115, 127
Active galaxies, 111, 127
Angular momentum, 47–48, 71, 76, 127
    conservation of, 128
Antimatter, 80, 118, 127
Antiparticle, 80, 127
Atoms, 15, 16, 18, 127
Axially symmetrical, 127
Axis, 47, 127

Bartel, Norbert, 112, 114, 115
Bekenstein, Jacob, 78
Bell, Jocelyn, 90–91
Bender, 101–102, 128
Big Bang, 125, 128
Big Crunch, 126, 128
Binary system, 74, 93, 96, 128
Birth of a star, 13–15

Blue shift, 58
Brown dwarfs, 103, 128
Black hole, 12, 29, 30, 34
    "an enormous nothing," 69–89
    as a lens, 70–71
    birth of, 45
    description, 128
    final stage of star, 13, 18, 21, 23, 25, 35
    journey to, 43–68
    star collapses, 40
    when star becomes one, 37
Blister, 120
Burnout, life cycle of a star, 13–27

Carbon atoms, 17
Carroll, Lewis G., 28
Centrifugal force, 99, 108
Chandrasekhar limit, 22, 23, 45, 83, 94, 128

Chandrasekhar, Subrahman-
yan, 22, 23
Charge, 74
Clumping, 13
Clusters of galaxies, 102
Cold star, 18, 22
Collapse, of a star, 55–59, 67
in slow motion, 65
Comet, 118
Companions, invisible, 93–99
Conservation of energy, 128
Curvature of spacetime, 36–
37, 38, 39, 124–125, 128
Cygnus A, 113
Cygnus X-1, 94, 95, 96, 97,
98, 99, 108, 128

Dark matter, 99–104, 128
Dimensions, 27
Dog Star, 19
Doppler effect, 50, 51, 94, 96,
110, 128
Dwarf galaxies, 103

Earth
as is, 29
shrunk to black hole, 31
shrunk to half present
radius, 30
Einstein, Albert, 25, 34, 39,
43, 51, 71, 83–84, 90
time, 59
Einstein Observatory, 97
Einstein ring, 71, 73, 128–129
Electromagnetic force, 129
Electromagnetic radiation, 129
Electromagnetic spectrum, 36,
50, 129
Electromagnetic waves, 34,
35, 61
Electron, 20, 103, 129
Elliptical orbits, 34

End of space, 39
Entropy, 77–78, 129
Ergosphere, 48–49, 77, 88,
129
Escape velocity, 29, 36–37,
39, 129
Event horizon, 38, 46, 71, 78,
81, 92
antiparticle pairs, 80
time, 63
what it is, 39–40, 129
Evidence, for black holes, 90–
115
Exclusion principle, 18, 20,
21, 23, 25, 26, 129
Experiments, 84–89

Fourth dimension, 27
Free-fall, 45–46, 129
Frequency, 61

Galactic core, 104–106
Galaxies, categories, 111
Gamma rays, 35, 86, 130
Garbage recycler, 120–122
Gravitational collapse, 45
Gravitational radiation, 54–55,
130
Gravitational radius, 130
Gravitational redshift, 50–55,
130
Graviton, 80, 130
Gravity, 15, 16, 17, 20, 25,
124–126, 130
bending of light, 31–35
capture of light, 35–42
exclusion principle, 20, 21
hazards of, 49–50
power of, 28–42
Gravity waves, 50, 51, 85, 130
Great Nebula, 14
Guth, Alan, 120

140

Hawking, Stephen, 78, 79, 81, 83, 90, 95, 119–120, 122
  wormhole, 68
Hawking radiation, 79, 82, 83, 88, 130
  and unmatched socks, 119–120
Helium, 15
Helium atoms, 17, 130
Hewish, Tony, 91
Hydrogen, 15, 16
Hydrogen atoms, 15, 17, 130

Intergalactic medium, 112
Invisible companions, 93–99
Ionized gas, 106, 131

Laplace, Pierre-Simon, 13
Lensing effect, 76, 130
LGM (Little Green Men), 91
Light
  at edge of spacetime, 108–115
  bending of, 31–35
  captive of, 35–42
Light waves, 61, 62
Little Green Men, 91

Magnetism, 99
Mass, 21, 23, 24, 35, 38, 71, 74, 130
Matter, 130–131
Matter-eater, 119
Matter particles, 25, 80
Matthews, Thomas, 110
Measuring, black hole, 71–84
  angular momentum, 71, 74, 76
  charge, 74
  chemical makeup of star, 74
  entropy, 77–78

mass, 71, 74, 76
rate of rotation, 76–77
size, 76
temperature, 78, 80–81, 83–84
Meteorites, 118
Milky Way, galaxy, 99, 104, 105, 111
Milne, A. A., 69
Misner, Charles, 120
Moon, 45
M87, galaxy, 100

Nebula, 109–110, 131
Neutrinos, 103
Neutron star, 18, 21, 23, 45, 90, 93, 94, 99, 131
  discovery of, 91
Neutron, 23, 131
Nuclear fusion, 15, 16, 131
Nuclei, 15
Nucleus, 131

Orion, 14

Particles, 18, 20, 34, 35, 36, 80, 131
  pairs, 131
  speed of light, 25
  wormholes, 68
Penrose process, 88, 122, 131
Penrose, Roger, 88
Photons, 31, 34, 35, 36, 37, 38, 40, 64, 67, 70, 80, 131
Plasma, ionized gas, 106, 131
Polar ring galaxies, 100, 131
Primordial black holes, 83, 103, 131
Proton, 15, 103, 132
Pulsars, 18, 91, 94, 132

Quantum state, 20, 132

141

Quasars, 109, 110–111, 112, 132

Radio astronomy, 100
Radio receivers, 11
Radio signals, 35
Radio waves, 35, 61, 91, 112
Radius, 132
Redshift, 50–55, 58, 110, 132
    time dilation, 60–63
Relativity, 92, 129, 132
Rotation, hazards of, 47–49

Sagittarius, constellation, 104, 105
Sandage, Alan, 110
Schmidt, Maarten, 110
Schwarzschild black hole, 132
Shapes, of black holes, 77
Shrunken star, 21
Singularity, 39, 42, 56, 59, 60, 67, 68, 70, 71, 92, 124, 126
    what it is, 40, 132
Sirius A (Dog Star), 18, 19
Sirius B, white dwarf, 19
Skeptics, 92
Solar mass, 23, 74, 132
Sound waves, 50
Spacetime, 25, 34
    what it is, 27, 132
Spacetime curvature, 32, 36–37, 38, 50, 67
    lights, 108–115
Speed of light, 25, 32, 36, 39, 47, 83, 109, 132
Spiral galaxies, 108
Star
    adult years, 16–17
    birth of, 13, 15
    life cycle, 13–27
    old age, 13, 17–26
Static limit, 48–49, 132

Stationary limit, 48, 132
Stony Tunguska River, 117–119
Sun, lifetime of, 17
Supernova, 23, 132

Telescopes, 11, 35
Temperature, of a black hole, 78–84
Theoretical physicists, 43
Theory of relativity, 92, 129, 132
Thorne, Kip, 95, 120
3C84, galaxy, 112, 114
Tidal effects, 45–47, 58, 68, 70, 89, 108, 132
    moon, 45
Time, 58–63
    Einstein, 59
    fourth dimension, 27
Time dilation, 59, 60–63, 89, 132
Time line, 55–57, 133
Treacherous voyage, 43–68
Tunguska, 117–119

Universal expansion, 125

Van den Bergh, Sidney, 100
Velocity, 20, 25
"Virtual" particles, 80, 81, 133
Visible spectrum, 35, 133

Wavelengths, 50, 53, 61, 81, 133
Wheeler, John, 12, 74, 75, 120, 125
White dwarf, 18, 20–21, 23, 93, 95, 133
    Sirius B, 19
Wild ideas, 116–123
    close encounter, 117–119

Wild ideas (continued)
    creation in the basement, 120
    donkey and carrot, 121
    Galactic Grand Prix, 122–123
    garbage recycler, 120

matter-eater, 119
unmatched socks, 119–120
Wimps (weakly interacting massive particles), 104
Wormhole, 67–68, 133

X rays, 35, 98, 133

ABOUT THE AUTHOR

Kitty Ferguson received her bachelor's and master's degrees from the Juilliard School of Music and has had a long and successful career as a professional musician. However, after a year spent with her husband (a Rutgers University professor) and family at Cambridge University, England, she decided to change course and devote herself full-time to a second lifelong interest, the physics of the universe—and especially to interpreting that subject for young readers. Her research has been wide ranging and included several conversations with Hawking and other experts such as John A. Wheeler (who coined the term *black hole*). Kitty Ferguson has two grown sons and a young daughter. Her home is in Chester, New Jersey.